Astronomy and Astrophysics Series

16

D0705456

The Foundations of Celestial Mechanics

Astronomy and Astrophysics Series

General Editor: A. G. Pacholczyk

1. T. L. Swihart
    Basic Physics of Stellar Atmospheres

2. T. L. Swihart
    Physics of Stellar Interiors

3. R. J. Weymann, T. L. Swihart, R. E. Williams, W. J. Cocke,
    A. G. Pacholczyk, J. E. Felten
    Lecture Notes on Introductory Theoretical Astrophysics

4. E. R. Craine
    A Handbook of Quasistellar and BL Lacertae Objects

5. A. G. Pacholczyk
    A Handbook of Radio Sources, Part 1

6. V. N. Zharkov, V. P. Trubitsyn
    Planetary Interiors

7. G. W. Collins, II
    The Virial Theorem in Stellar Astrophysics

8. G. S. Rossano, E. R. Craine
    Near Infrared Photographic Sky Survey: A Field Index

9. D. J. Raine, M. Heller
    The Science of Space-Time

10. G. R. Gisler, E. D. Friel
    Index of Galaxy Spectra

11. Z. K. Alksne, Ya. Ya. Ikaunieks
    Carbon Stars

12. T. L. Swihart
    Radiation Transfer and Stellar Atmospheres

14. M. Heller
    Questions to the Universe

16. G. W. Collins, II
    The Foundations of Celestial Mechanics

The Foundations of Celestial Mechanics

by

George W. Collins, II

The Ohio State University

Pachart Publishing House

Tucson

G. W. Collins, II (1937 -   )
The Foundations of Celestial Mechanics

Library of Congress Catalog Card Number: 88-061797
International Standard Book Number: 0-88126-009-6

Pachart Astronomy and Astrophysics Series Volume 16

Pachart Publishing House
A Division of Pachart Foundation
A Nonprofit Association
1130 San Lucas Circle
P. O. Box 35549
Tucson, Arizona 85704

To C.M.Huffer, who taught it the old way,
but who cared that we learn.

List of Figures

Table of Contents

Preface

This book resulted largely from an accident. I was faced with teaching celestial mechanics at The Ohio State University during the Winter Quarter of 1988. As a result of a variety of errors, no textbook would be available to the students until very late in the quarter at the earliest. Since my approach to the subject has generally been non-traditional, a textbook would have been of marginal utility in any event, so I decided to write up what I would be teaching so that the students would have something to review beside lecture notes. This is the result.

Celestial mechanics is a course that is fast disappearing from the curricula of astronomy departments across the country. The pressure to present the new and exciting discoveries of the past quarter century has led to the demise of a number of traditional subjects. In point of fact, very few astronomers are involved in traditional celestial mechanics. Indeed, I doubt if many could determine the orbital elements of a passing comet and predict its future path based on three positional measurements without a good deal of study. This was a classical problem in celestial mechanics at the turn of this century and any astronomer worth his degree would have had little difficulty solving it. Times, as well as disciplines, change and I would be among the first to recommend the deletion from the college curriculum of the traditional course in celestial mechanics such as the one I had twenty five years ago.

There are, however, many aspects of celestial mechanics that are common to other disciplines of science. A knowledge of the mathematics of coordinate transformations will serve well any astronomer, whether observer or theoretician. The classical mechanics of Lagrange and Hamilton will prove useful to anyone who must sometime in a career analyze the dynamical motion of a planet, star, or galaxy. It can also be used to arrive at the equations of motion for objects in the solar system. The fundamental constraints on the N-body problem should be familiar to anyone who would hope to understand the dynamics of stellar systems. And perturbation theory is one of the most widely used tools in theoretical physics. The fact that it is more successful in quantum mechanics than in celestial mechanics speaks more to the relative intrinsic difficulty of the theories than to the methods. Thus celestial mechanics can be used as a vehicle to introduce students to a whole host of subjects that they should know. I feel that this is perhaps the appropriate role for the contemporary study of celestial mechanics at the undergraduate level.

This is not to imply that there are no interesting problems left in celestial mechanics. There still exists no satisfactory explanation for the Kirkwood Gaps of the asteroid belt. The ring system of Saturn is still far from understood. The theory of the motion of the moon may give us clues as to the origin of the moon, but the issue is still far from resolved. Unsolved problems are simply too hard for solutions to be found by any who do not devote a great deal of time and effort to them. An introductory course cannot hope to prepare students adequately to tackle these problems. In addition, many of the traditional approaches to problems were developed to minimize computation by accepting only approximate solutions. These approaches are truly fossils of interest only to those who study the development and history of science. The computational power available to the contemporary scientist enables a more straightforward, though perhaps less elegant, solution to many of the traditional problems of celestial mechanics. A student interested in the contemporary approach to such problems would be well advised to obtain a through grounding in the numerical solution of differential equations before approaching these problems of celestial mechanics.

I have mentioned a number of areas of mathematics and physics that bear on the study of celestial mechanics and suggested that it can provide examples for the application of these techniques to practical problems. I have attempted to supply only an introduction to these subjects. The reader should not be disappointed that these subjects are not covered completely and with full rigor as this was not my intention. Hopefully, his or her appetite will be 'whetted' to learn more as each constitutes a significant course of study in and of itself. I hope that the reader will find some unity in the application of so many diverse fields of study to a single subject, for that is the nature of the study of physical science. In addition, I can only hope that some useful understanding relating to celestial mechanics will also be conveyed. In the unlikely event that some students will be called upon someday to determine the ephemeris of a comet or planet, I can only hope that they will at least know how to proceed.

As is generally the case with any book, many besides the author take part in generating the final product. Let me thank Peter Stoycheff and Jason Weisgerber for their professional rendering of my pathetic drawings and Ryland Truax for reading the manuscript. In addition, Jason Weisgerber carefully proof read the final copy of the manuscript finding numerous errors that evaded my impatient eyes. Special thanks are due Elizabeth Roemer of the Steward Observatory for carefully reading the manuscript and catching a large number of embarrassing errors and generally improving the result. Those errors that remain are clearly my responsibility and I sincerely hope that they are not too numerous and distracting.

George W. Collins, II

June 24, 1988

1

Introduction and Mathematics Review

1.1 The Nature of Celestial Mechanics

Celestial mechanics has a long and venerable history as a discipline. It would be fair to say that it was the first area of physical science to emerge from Newton's theory of mechanics and gravitation put forth in the Principia. It was Newton's ability to describe accurately the motion of the planets under the concept of a single universal set of laws that led to his fame in the seventeenth century. The application of Newtonian mechanics to planetary motion was honed to so fine an edge during the next two centuries that by the advent of the twentieth century the description of planetary motion was refined enough that the departure of prediction from observation by 43 arcsec in the precession of the perihelion of Mercury's orbit was a major factor in the replacement of Newton's theory of gravity by the General Theory of Relativity.

At the turn of the century no professional astronomer would have been considered properly educated if he could not determine the location of a planet in the local sky given the orbital elements of that planet. The reverse would also have been expected. That is, given three or more positions of the planet in the sky for three different dates, he should be able to determine the orbital elements of that planet preferably in several ways. It is reasonably safe to say that few contemporary astronomers could accomplish this without considerable study. The emphasis of astronomy has shifted dramatically during the past fifty years. The techniques of classical celestial mechanics developed by Gauss, Lagrange, Euler and many others have more or less been consigned to the

1

history books. Even in the situation where the orbits of spacecraft are required, the accuracy demanded is such that much more complicated mechanics is necessary than for planetary motion, and these problems tend to be dealt with by techniques suited to modern computers.

However, the foundations of classical celestial mechanics contains elements of modern physics that should be understood by every physical scientist. It is the understanding of these elements that will form the primary aim of the book while their application to celestial mechanics will be incidental. A mastery of these fundamentals will enable the student to perform those tasks required of an astronomer at the turn of the century and also equip him to deal with more complicated problems in many other fields.

The traditional approach to celestial mechanics well into the twentieth century was incredibly narrow and encumbered with an unwieldy notation that tended to confound rather than elucidate. It wasn't until the 1950s that vector notation was even introduced into the subject at the textbook level. Since throughout this book we shall use the now familiar vector notation along with the broader view of classical mechanics and linear algebra, it is appropriate that we begin with a review of some of these concepts.

1.2 Scalars, Vectors, Tensors, Matrices and Their Products

While most students of the physical sciences have encountered scalars and vectors throughout their college career, few have had much to do with tensors and fewer still have considered the relations between these concepts. Instead they are regarded as separate entities to be used under separate and specific conditions. Other students regard tensors as the unfathomable language of General Relativity and therefore comprehensible only to the intellectually elite. This latter situation is unfortunate since tensors are incredibly useful in the wide range of modern theoretical physics and the sooner one vanquishes his fear of them the better. Thus, while we won't make great use of them in this book, we will introduce them and describe their relationship to vectors and scalars.

a. Scalars

The notion of a scalar is familiar to anyone who has completed a freshman course in physics. A single number or symbol is used to describe some physical quantity. In truth, as any mathematician will tell you, it is not necessary for the

2

scalar to represent anything physical. But since this is a book about physical science we shall narrow our view to the physical world. There is, however, an area of mathematics that does provide a basis for defining scalars, vectors, etc. That area is set theory and its more specialized counterpart, group theory. For a collection or set of objects to form a group there must be certain relations between the elements of the set. Specifically, there must be a "Law" which describes the result of "combining" two members of the set. Such a "Law" could be *addition*. Now if the action of the law upon any two members of the set produces a third member of the set, the set is said to be "closed" with respect to that law. If the set contains an element which, when combined under the law with any other member of the set, yields that member unchanged, that element is said to be the identity element. Finally, if the set contains elements which are inverses, so that the combination of a member of the set with its inverse under the "Law" yields the identity element, then the set is said to form a group under the "Law".

The integers (positive and negative, including zero) form a group under addition. In this instance, the identity element is zero and the operation that generates inverses is subtraction so that the negative integers represent the inverse elements of the positive integers. However, they do not form a group under multiplication as each inverse is a fraction. On the other hand the rational numbers do form a group under both addition and multiplication. Here the identity element for addition is again zero, but under multiplication it is one. The same is true for the real and complex numbers. Groups have certain nice properties; thus it is useful to know if the set of objects forms a group or not. Since scalars are generally used to represent real or complex numbers in the physical world, it is nice to know that they will form a group under multiplication and addition so that the inverse operations of subtraction and division are defined. With that notion alone one can develop all of algebra and calculus which are so useful in describing the physical world. However, the notion of a vector is also useful for describing the physical world and we shall now look at their relation to scalars.

b. Vectors

A vector has been defined as "an ordered n-tuple of numbers". Most find that this technically correct definition needs some explanation. There are some physical quantities that require more than a single number to fully describe them. Perhaps the most obvious is an object's location in space. Here we require three numbers to define its location (four if we

3

include time). If we insist that the order of those three
numbers be the same, then we can represent them by a single
symbol called a vector. In general, vectors need not be limited
to three numbers; one may use as many as is necessary to
characterize the quantity. However, it would be useful if the
vectors also formed a group and for this we need some "Laws"
for which the group is closed. Again addition and
multiplication seem to be the logical laws to impose. Certainly
vector addition satisfies the group condition, namely that the
application of the "law" produces an element of the set. The
identity element is a 'zero-vector' whose components are all
zero. However, the commonly defined "laws" of multiplication do
not satisfy this condition.

Consider the vector scalar product, also known as the
inner product, which is defined as

$$\vec{A}.\vec{B} = c \equiv \sum_i A_i B_i \quad . \tag{1.2.1}$$

Here the result is a scalar which is clearly a different type
of quantity than a vector. Now consider the other well known
'vector product', sometimes called the cross product, which in
ordinary cartesian coordinates is defined as

$$\vec{A} \times \vec{B} \equiv \begin{vmatrix} \hat{i} & \hat{j} & \hat{k} \\ A_i & A_j & A_k \\ B_i & B_j & B_k \end{vmatrix} = \hat{i}(A_j B_k - A_k B_j) - \hat{j}(A_i B_k - A_k B_i)$$
$$+ \hat{k}(A_i B_j - A_j B_i) \quad . \tag{1.2.2}$$

This appears to satisfy the condition that the result is a
vector. However as we shall see, the vector produced by this
operation does not behave in the way in which we would like all
vectors to behave.

Finally, there is a product law known as the tensor, or
outer product that is useful to define as

$$\left. \begin{array}{l} \vec{A}\vec{B} \equiv C , \\[2ex] C_{ij} = A_i B_j \end{array} \right\} \quad . \tag{1.2.3}$$

Here the result of applying the "law" is an ordered array of
n×m numbers where n and m are the dimensionalities of the
vectors $\vec{A}$ and $\vec{B}$ respectively. Such a result is clearly not a
vector and so vectors under this law do not form a group. In
order to provide a broader concept wherein we can understand
scalars and vectors as well as the results of the outer
product, let us briefly consider the quantities knows as
tensors.

4

c. Tensors and Matrices

In general a tensor has N^n components or elements. N is known as the dimensionality of the tensor by analogy with the notion of a vector while n is called the rank. Thus vectors are simply tensors of rank unity while scalars are tensors of rank zero. If we consider the set of all tensors, then they form a group under addition and all of the vector products. Indeed the inner product can be generalized for tensors of rank m and n. The result will be a tensor of rank $|m-n|$. Similarly the outer product can be so defined that the outer product of tensors with rank m and n is a tensor of rank $|m+n|$.

One obvious way of representing tensors of rank two is by denoting them as matrices. Thus the arranging of the N^2 components in an N×N array will produce the familiar square matrix. The scalar product of a matrix and vector should then yield a vector by

$$\left. \begin{array}{l} A.\vec{B} = \vec{C} \quad , \\[2mm] C_i = \sum_j A_{ij} B_j \end{array} \right\} \quad , \tag{1.2.4}$$

while the outer product would result in a tensor of rank three from

$$\left. \begin{array}{l} A\vec{B} = \underline{C} \quad , \\[2mm] C_{ijk} = A_{ij} B_k \end{array} \right\} \quad . \tag{1.2.5}$$

An important tensor of rank two is called the unit tensor whose elements are the Kronecker delta and for two dimensions is written as

$$1 = \begin{pmatrix} 1 & 0 \\ 0 & 1 \end{pmatrix} = \delta_{ij} \quad . \tag{1.2.6}$$

Clearly the scalar product of this tensor with a vector yields the vector itself. There is a parallel tensor of rank three known as the Levi-Civita tensor (or more correctly tensor density) which is a three index tensor whose elements are zero when any two indices are equal. When the indices are all different the value is +1 or -1 depending on whether the index sequence can be obtained by an even or odd permutation of 1,2,3 respectively. Thus the elements of the Levi-Civita tensor can be written in terms of three matrices as

$$\epsilon_{1jk} = \begin{pmatrix} 0 & 0 & 0 \\ 0 & 0 & +1 \\ 0 & -1 & 0 \end{pmatrix} \; ,$$

$$\epsilon_{2jk} = \begin{pmatrix} 0 & 0 & -1 \\ 0 & 0 & 0 \\ +1 & 0 & 0 \end{pmatrix} \; , \qquad \Biggr\} \qquad \text{(1.2.7)}$$

$$\epsilon_{3jk} = \begin{pmatrix} 0 & -1 & 0 \\ +1 & 0 & 0 \\ 0 & 0 & 0 \end{pmatrix} \; .$$

One of the utilities of this tensor is that it can be used to express the vector cross product as follows

$$\vec{A} \times \vec{B} = \underline{\epsilon} \cdot (\vec{A}\vec{B}) = \sum_{jk} \epsilon_{ijk} A_j B_k = C_i \; . \qquad \text{(1.2.8)}$$

As we shall see later, while the rule for calculating the rank correctly implies that the vector cross product as expressed by equation (1.2.8) will yield a vector, there are reasons for distinguishing between this type of vector and the normal vectors $\vec{A}$ and $\vec{B}$. These same reasons extend to the correct naming of the Levi-Civita tensor as the Levi-Civita tensor density. However, before this distinction can be made clear, we shall have to understand more about coordinate transformations and the behavior of both vectors and tensors that are subject to them.

The normal matrix product is certainly different from the scalar or outer product and serves as an additional multiplication "law" for second rank tensors. The standard definition of the matrix product is

$$\begin{aligned} AB &= C \; , \\ C_{ij} &= \sum A_{ik} B_{kj} \end{aligned} \Biggr\} \; . \qquad \text{(1.2.9)}$$

Only if the matrices can be resolved into the outer product of two vectors so that

$$\begin{aligned} A &= \vec{a}\vec{\alpha} \\ B &= \vec{\beta}\vec{b} \end{aligned} \Biggr\} \; , \qquad \text{(1.2.10)}$$

can the matrix product be written in terms of the products that we have already defined - namely

$$AB = \vec{a}\vec{b}(\vec{\alpha} \cdot \vec{\beta}) \; . \qquad \text{(1.2.11)}$$

6

There is much more that can, and perhaps should, be said about matrices. Indeed, entire books have been written about their properties. However, we shall consider only some of those properties within the notion of a group. Clearly the unit tensor (or unit matrix) given by equation (1.2.6) represents the unit element of the matrix group under matrix multiplication. The unit under addition is simply a matrix whose elements are all zero, since matrix addition is defined by

$$
\left.
\begin{array}{l}
A + B = C \\
\\
A_{ij} + B_{ij} = C_{ij}
\end{array}
\right\}
\qquad (1.2.12)
$$

Remember that the unit element of any group forms the definition of the inverse element. Clearly the inverse of a matrix *under addition* will simply be that matrix whose elements are the negative of the original matrix, so that their sum is zero. However, the inverse of a matrix under matrix multiplication is quite another matter. We can certainly define the process by

$$
AA^{-1} = 1 \quad , \qquad (1.2.13)
$$

but the process by which A^{-1} is actually computed is lengthy and beyond the scope of this book. We can further define other properties of a matrix such as the *transpose* and the *determinant*. The transpose of a matrix A with elements A_{ij} is just

$$
A^T \equiv A_{ji} \quad , \qquad (1.2.14)
$$

while the determinant is obtained by expanding the matrix by minors as is done in Kramer's rule for the solution of linear algebraic equations. For a 3×3 matrix, this would become

$$
\det A = \det \begin{vmatrix} a_{11} & a_{12} & a_{13} \\ a_{21} & a_{22} & a_{23} \\ a_{31} & a_{32} & a_{33} \end{vmatrix} = \begin{aligned} &+ a_{11}(a_{22}a_{33} - a_{23}a_{32}) \\ &- a_{12}(a_{21}a_{33} - a_{23}a_{31}) \\ &+ a_{13}(a_{21}a_{32} - a_{22}a_{31}) \quad . \end{aligned} \qquad (1.2.15)
$$

The matrix is said to be symmetric if $A_{ij} = A_{ji}$. Finally, if the matrix elements are complex so that the transpose element is the complex conjugate of its counterpart, the matrix is said to be Hermitian. Thus for a Hermitian matrix H the elements obey

$$H_{ij} = \bar{H}_{ji} \quad , \tag{1.2.16}$$

where $\bar{H}_{ji}$ is the complex conjugate of H_{ij}.

1.3 Commutativity, Associativity, and Distributivity

Any "law" that is defined on the elements of a set may have certain properties that are important for the implementation of that "law" and the resultant elements. For the sake of generality, let us denote the "law" by $\wedge$, which can stand for any of the products that we have defined. Now any such law is said to be commutative if

$$A \wedge B = B \wedge A \quad . \tag{1.3.1}$$

Of all the laws we have discussed only addition and the scalar product are commutative. This means that considerable care must be observed when using the outer, vector-cross, or matrix products, as the order in which terms appear in a product will make a difference in the result.

Associativity is a somewhat weaker condition and is said to hold for any law when

$$(A \wedge B) \wedge C = A \wedge (B \wedge C) \quad . \tag{1.3.2}$$

In other words the order in which the law is applied to a string of elements doesn't matter if the law is associative. Here addition, the scalar, and matrix products are associative while the vector cross product and outer product are, in general, not.

Finally, the notion of distributivity involves the relation between two different laws. These are usually addition and one of the products. Our general purpose law $\wedge$ is said to be distributive with respect to addition if

$$A \wedge (B+C) = (A \wedge B) + (A \wedge C) \quad . \tag{1.3.3}$$

This is usually the weakest of all conditions on a law and here all of the products we have defined pass the test. They are all distributive with respect to addition. The main function of remembering the properties of these various products is to insure that mathematical manipulations on expressions involving them are done correctly.

8

1.4 Operators

The notion of operators is extremely important in mathematical physics and there are entire books written on the subject. Most students usually first encounter operators in calculus when the notation [d/dx] is introduced to denote the derivative of a function. In this instance the operator stands for taking the limit of the difference between adjacent values of some function of x divided by the difference between the adjacent values of x as that difference tends toward zero. This is a fairly complicated set of instructions represented by a relatively simple set of symbols. The designation of some symbol to represent a collection of operations is said to represent the definition of an operator. Depending on the details of the definition, the operators can often be treated as if they were quantities and subjected to algebraic manipulations. The extent to which this is possible is determined by how well the operators satisfy the conditions for the group on which the algebra or mathematical system in question is defined.

We shall make use of a number of operators in this book, the most common of which is the "del" operator or "nabla". It is usually denoted by the symbol ∇ and is a vector operator defined in cartesian coordinates as

$$\nabla \equiv \hat{i}\frac{\partial}{\partial x} + \hat{j}\frac{\partial}{\partial y} + \hat{k}\frac{\partial}{\partial z} \qquad (1.4.1)$$

This single operator, when combined with the some of the products defined above, constitutes the foundation of vector calculus. Thus the divergence, gradient, and curl are defined as

$$\left. \begin{array}{l} \nabla \cdot \vec{A} = \beta \\[2mm] \nabla \alpha = \vec{B} \\[2mm] \nabla \times \vec{A} = \vec{C} \end{array} \right\} \qquad (1.4.2)$$

respectively. If we consider $\vec{A}$ to be a continuous vector function of the independent variables that make up the space in which it is defined, then we may give a physical interpretation for both the divergence and curl. The divergence of a vector field is a measure of the amount that the field spreads or contracts at some given point in the space (see Figure 1.1).

9

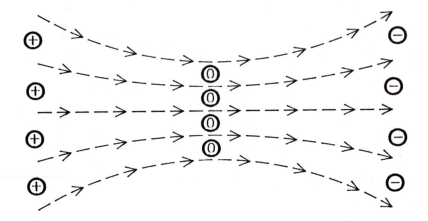

Figure 1.1 schematically shows the divergence of
a vector field. In the region where the arrows of
the vector field converge, the divergence is
positive, implying an increase in the source of
the vector field. The opposite is true for the
region where the field vectors diverge.

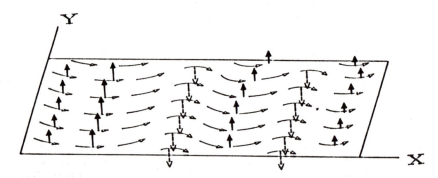

Figure 1.2 schematically shows the curl of a
vector field. The direction of the curl is
determined by the "right hand rule" while the
magnitude depends on the rate of change of the x-
and y-components of the vector field with respect
to y and x.

10

The curl is somewhat harder to visualize. In some sense it represents the amount that the field rotates about a given point. Some have called it a measure of the "swirliness" of the field. If in the vicinity of some point in the field, the vectors tend to veer to the left rather than to the right, then the curl will be a vector pointing up normal to the net rotation with a magnitude that measures the degree of rotation (see Figure 1.2). Finally, the gradient of a scalar field is simply a measure of the direction and magnitude of the maximum rate of change of that scalar field (see Figure 1.3).

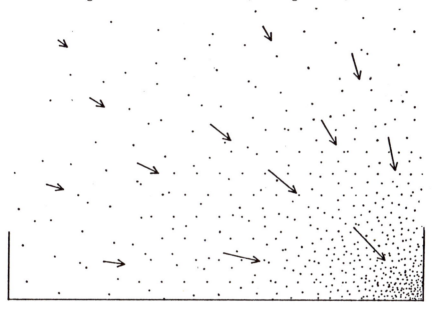

Figure 1.3 schematically shows the gradient of the scalar dot-density in the form of a number of vectors at randomly chosen points in the scalar field. The direction of the gradient points in the direction of maximum increase of the dot-density while the magnitude of the vector indicates the rate of change of that density.

With these simple pictures in mind it is possible to generalize the notion of the Del-operator to other quantities. Consider the gradient of a vector field. This represents the outer product of the Del-operator with a vector. While one doesn't see such a thing often in freshman physics, it does occur in more advanced descriptions of fluid mechanics (and many other places). We now know enough to understand that the result of this operation will be a tensor of rank two which we

11

can represent as a matrix. What do the components mean? Generalize from the scalar case. The nine elements of the vector gradient can be viewed as three vectors denoting the direction of the maximum rate of change of *each* of the components of the original vector. The nine elements represent a perfectly well defined quantity and it has a useful purpose in describing many physical situations. One can also consider the divergence of a second rank tensor, which is clearly a vector. In hydrodynamics, the divergence of the pressure tensor may reduce to the gradient of the scalar gas pressure if the macroscopic flow of the material is small compared to the internal speed of the particles that make up the material.

Thus by combining the various products defined in this chapter with the familiar notions of vector calculus, we can formulate a much richer description of the physical world. This review of scalar and vector mathematics along with the all-too-brief introduction to tensors and matrices will be useful, not only in the development of celestial mechanics, but in the general description of the physical world. However, there is another broad area of mathematics on which we must spend some time. To describe events in the physical world, it is common to frame them within some system of coordinates. We will now consider some of these coordinates and the transformations between them.

a. Common Del-Operators

Cylindrical Coordinates

Orthogonal Line Elements

dr, $rd\vartheta$, dz

Divergence

$$\nabla \cdot \vec{A} = \frac{1}{r}\frac{\partial(rA_r)}{\partial r} + \frac{1}{r}\frac{\partial A_\vartheta}{\partial\vartheta} + \frac{\partial A_z}{\partial z}$$

Components of the Gradient

$$(\nabla a)_r = \frac{\partial a}{\partial r}$$

$$(\nabla a)_\vartheta = \frac{1}{r}\frac{\partial a}{\partial\vartheta}$$

$$(\nabla a)_z = \frac{\partial a}{\partial z}$$

Components of the Curl

$$(\nabla\times\vec{A})_r = \frac{1}{r}\frac{\partial A_z}{\partial\vartheta} - \frac{\partial A_\vartheta}{\partial z}$$

$$(\nabla\times\vec{A})_\vartheta = \frac{\partial A_r}{\partial z} - \frac{\partial A_z}{\partial r}$$

$$(\nabla\times\vec{A})_z = \frac{1}{r}\left[\frac{\partial(rA_\vartheta)}{\partial r} - \frac{\partial A_r}{\partial\vartheta}\right]$$

Spherical Coordinates

Orthogonal Line Elements

dr, $rd\theta$, $r\sin\theta\, d\phi$

Divergence

$$\nabla \cdot \vec{A} = \frac{1}{r^2}\frac{\partial(r^2 A_r)}{\partial r}$$

$$+ \frac{1}{r\sin\theta}\frac{\partial(A_\theta\sin\theta)}{\partial\theta} + \frac{1}{r\sin\theta}\frac{\partial A_\phi}{\partial\phi}$$

Components of the Gradient

$$(\nabla a)_r = \frac{\partial a}{\partial r}$$

$$(\nabla a)_\theta = \frac{1}{r}\frac{\partial a}{\partial\theta}$$

$$(\nabla a)_\phi = \frac{1}{r\sin\theta}\frac{\partial a}{\partial\phi}$$

Components of the Curl

$$(\nabla\times\vec{A})_r = \frac{1}{r\sin\theta}\left[\frac{\partial(A_\phi\sin\theta)}{\partial\theta} - \frac{\partial A_\theta}{\partial\phi}\right]$$

$$(\nabla\times\vec{A})_\theta = \frac{1}{r\sin\theta}\frac{\partial A_r}{\partial\phi} - \frac{1}{r}\frac{\partial(rA_\phi)}{\partial r}$$

$$(\nabla\times\vec{A})_\phi = \frac{1}{r}\left[\frac{\partial(rA_\theta)}{\partial r} - \frac{\partial A_r}{\partial\theta}\right]$$

Chapter 1: Exercises

1. Find the components of the vector $\vec{A} = \nabla \times (\nabla \times \vec{B})$ in spherical coordinates.

2. Show that:

    a. $\nabla \times (\vec{A} \times \vec{B}) = (B.\nabla)\vec{A} - \vec{B}(\nabla.\vec{A}) - (\vec{A}.\nabla)\vec{B} + \vec{A}(\nabla.\vec{B})$.

    b. $\nabla.(\vec{A} \times \vec{B}) = \vec{B}.(\nabla \times \vec{A}) - \vec{A}.(\nabla \times \vec{B})$.

3. Show that:

    $\nabla(\vec{A}.\vec{B}) = (\vec{B}.\nabla)\vec{A} + (\vec{A}.\nabla)\vec{B} + \vec{B} \times (\nabla \times \vec{A}) + \vec{A} \times (\nabla \times \vec{B})$.

4. If T is a tensor of rank 2 with components T_{ij}, show that $\nabla.T$ is a vector and find the components of that vector.

Useful Vector Identities

$$\nabla.(a\vec{A}) = a\nabla.\vec{A} + \vec{A}.\nabla a \quad . \tag{a1}$$

$$\nabla \times (\alpha\vec{A}) = a\nabla \times \vec{A} + \vec{A} \times \nabla a \quad . \tag{a2}$$

$$\nabla \times (\nabla \times \vec{A}) = \nabla(\nabla.\vec{A}) - \nabla.(\nabla\vec{A}) \quad . \tag{a3}$$

$$\nabla \times (\vec{A} \times \vec{B}) = (B.\nabla)\vec{A} - \vec{B}(\nabla.\vec{A}) - (\vec{A}.\nabla)\vec{B} + \vec{A}(\nabla.\vec{B}) \quad . \tag{a4}$$

$$\nabla.(\vec{A} \times \vec{B}) = \vec{B}.(\nabla \times \vec{A}) - \vec{A}.(\nabla \times \vec{B}) \quad . \tag{a5}$$

$$\nabla(\vec{A}.\vec{B}) = (\vec{B}.\nabla)\vec{A} + (\vec{A}.\nabla)\vec{B} + \vec{B} \times (\nabla \times \vec{A}) + \vec{A} \times (\nabla \times \vec{B}) \quad . \tag{a6}$$

$$\nabla.(\nabla\alpha) \equiv \nabla^2\alpha = \text{Laplacian of } \alpha \ . \tag{a7}$$

In cartesian coordinates:

$$(\vec{A}.\nabla)\vec{B} = \hat{i}\left[A_x\frac{\partial B_x}{\partial x} + A_y\frac{\partial B_x}{\partial y} + A_z\frac{\partial B_x}{\partial z}\right] + \hat{j}\left[A_x\frac{\partial B_y}{\partial x} + A_y\frac{\partial B_y}{\partial y} + A_z\frac{\partial B_y}{\partial z}\right]$$

$$\hat{k}\left[A_x\frac{\partial B_z}{\partial x} + A_y\frac{\partial B_z}{\partial y} + A_z\frac{\partial B_z}{\partial z}\right] \quad . \tag{a8}$$

14

2

Coordinate Systems
and Coordinate Transformations

The field of mathematics known as topology describes space in a very general sort of way. Many spaces are exotic and have no counterpart in the physical world. Indeed, in the hierarchy of spaces defined within topology, those that can be described by a coordinate system are among the more sophisticated. These are the spaces of interest to the physical scientist because they are reminiscent of the physical space in which we exist. The utility of such spaces is derived from the presence of a coordinate system which allows one to describe phenomena that take place within the space. However, the spaces of interest need not simply be the physical space of the real world. One can imagine the temperature-pressure-density space of thermodynamics or many of the other spaces where the dimensions are physical variables. One of the most important of these spaces for mechanics is phase space. This is a multi-dimensional space that contains the position and momentum coordinates for a collection of particles. Thus physical spaces can have many forms. However, they all have one thing in common. They are described by some coordinate system or frame of reference. Imagine a set of rigid rods or vectors all connected at a point. Such a set of 'rods' is called a frame of reference. If every point in the space can uniquely be projected onto the rods so that a unique collection of rod-points identify the point in space, the reference frame is said to span the space.

2.1 Orthogonal Coordinate Systems

If the vectors that define the coordinate frame are locally perpendicular, the coordinate frame is said to be orthogonal. Imagine a set of unit basis vectors $\hat{e}_i$ that span some space. We can express the condition of orthogonality by

$$\hat{e}_i \cdot \hat{e}_j = \delta_{ij} \quad , \tag{2.1.1}$$

where δ_{ij} is the Kronecker delta that we introduced in the previous chapter. Such a set of basis vectors is said to be orthogonal and will span a space of n-dimensions where n is the number of vectors $\hat{e}_i$. It is worth noting that the space need not be euclidean. However, if the space is euclidean and the coordinate frame is orthogonal, then the coordinate frame is said to be a cartesian frame. The standard xyz coordinate frame is a cartesian frame. One can imagine such a coordinate frame drawn on a rubber sheet. If the sheet is distorted in such a manner that the local orthogonality conditions are still met, the coordinate frame may remain orthogonal but the space may no longer be a euclidean space. For example, consider the ordinary coordinates of latitude and longitude on the surface of the earth. These coordinates are indeed orthogonal but the surface is not the euclidean plane and the coordinates are not cartesian.

Of the orthogonal coordinate systems, there are several that are in common use for the description of the physical world. Certainly the most common is the cartesian or rectangular coordinate system (xyz). Probably the second most common and of paramount importance for astronomy is the system of spherical or polar coordinates (r, θ, ϕ). Less common but still very important are the cylindrical coordinates (r, ϑ, z). There is a total of thirteen orthogonal coordinate systems in which Laplace's equation is separable, and knowledge of their existence (see Morse and Feshback[1]) can be useful for solving problems in potential theory. Recently the dynamics of ellipsoidal galaxies has been understood in a semi-analytic manner by employing ellipsoidal coordinates and some potentials defined therein. While these more exotic coordinates were largely concerns of the nineteenth century mathematical physicists, they still have relevance today. Often the most important part of solving a problem in mathematical physics is the choice of the proper coordinate system in which to do the analysis.

In order to completely define any coordinate system one must do more than just specify the space and coordinate geometry. In addition, the origin of the coordinate system and its orientation must be given. In celestial mechanics there are three important locations for the origin. For observation, the

16

origin can be taken to be the observer (topocentric coordinates). However for interpretation of the observations it is usually necessary to refer the observations to coordinate systems with their origin at the center of the earth (geocentric coordinates) or the center of the sun (heliocentric coordinates) or at the center of mass of the solar system (barycentric coordinates). The orientation is only important when the coordinate frame is to be compared or transformed to another coordinate frame. This is usually done by defining the zero-point of some coordinate with respect to the coordinates of the other frame as well as specifying the relative orientation.

2.2 Astronomical Coordinate Systems

The coordinate systems of astronomical importance are nearly all spherical coordinate systems. The obvious reason for this is that most all astronomical objects are remote from the earth and so appear to move on the backdrop of the celestial sphere. While one may still use a spherical coordinate system for nearby objects, it may be necessary to choose the origin to be the observer to avoid problems with parallax. These orthogonal coordinate frames will differ only in the location of the origin and their relative orientation to one another. Since they have their foundation in observations made from the earth, their relative orientation is related to the orientation of the earth's rotation axis with respect to the stars and the sun. The most important of these coordinate systems is the Right Ascension - Declination coordinate system

a. The Right Ascension - Declination Coordinate System

This coordinate system is a spherical-polar coordinate system where the polar angle, instead of being measured from the axis of the coordinate system, is measured from the system's equatorial plane. Thus the declination is the angular complement of the polar angle. Simply put, it is the angular distance to the astronomical object measured north or south from the equator of the earth as projected out onto the celestial sphere. For measurements of distant objects made from the earth, the origin of the coordinate system can be taken to be at the center of the earth. At least the 'azimuthal' angle of the coordinate system is measured in the proper fashion. That is, if one points the thumb of his right hand toward the north pole, then the fingers will point in the direction of increasing Right Ascension. Some remember it by noting that the Right Ascension of rising or ascending stars increases with

17

time. There is a tendency for some to face south and think that the angle should increase to their right as if they were looking at a map. This is exactly the reverse of the true situation and the notion so confused airforce navigators during the second world war that the complementary angle, known as the sidereal hour angle, was invented. This angular coordinate is just 24 hours minus the Right Ascension.

Another aspect of this Right Ascension that many find confusing is that it is not measured in any common angular measure like degrees or radians. Rather it is measured in hours, minutes, and seconds of time. However, these units are the natural ones as the rotation of the earth on its axis causes any fixed point in the sky to return to the same place after about 24 hours.

We still have to define the zero-point from which the Right Ascension angle is measured. This also is inspired by the orientation of the earth. The projection of the orbital plane of the earth on the celestial sphere is described by the path taken by the sun during the year. This path is called the ecliptic. Since the rotation axis of the earth is inclined to the orbital plane, the ecliptic and equator, represented by great circles on the celestial sphere, cross at two points 180° apart. The points are known as equinoxes, for when the sun is at them it will lie in the plane of the equator of the earth and the length of night and day will be equal. The sun will visit each once a year, one when it is headed north along the ecliptic, and the other when it is headed south. The former is called the vernal equinox as it marks the beginning of spring in the northern hemisphere while the latter is called the autumnal equinox. The point in the sky known as the vernal equinox is the zero-point of the Right Ascension coordinate, and the Right Ascension of an astronomical object is measured eastward from that point in hours, minutes, and seconds of *time*.

While the origin of the coordinate system can be taken to be the center of the earth, it might also be taken to be the center of the sun. Here the coordinate system can be imagined as simply being shifted without changing its orientation until its origin corresponds with the center of the sun. Such a coordinate system is useful in the studies of stellar kinematics. For some studies in stellar dynamics, it is necessary to refer to a coordinate system with an orgin at the center of mass of the earth-moon system. These are known as barycentric coordinates. Indeed, since the term barycenter refers to the center of mass, the term barycentric coordinates may also be used to refer to a coordinate system whose origin is at the center of mass of the solar system. The domination of the sun over the solar system insures that this origin will be very near, but not the same as the origin of the heliocentric coordinate system. Small as the differences of origin between the heliocentric and barycentric coordinates is, it is large enough to be significant for some problems such as the timing of pulsars.

b. Ecliptic Coordinates

The ecliptic coordinate system is used largely for studies involving planets and asteroids as their motion, with some notable exceptions, is confined to the zodiac. Conceptually it is very similar to the Right Ascension-Declination coordinate system. The defining plane is the ecliptic instead of the equator and the "azimuthal" coordinate is measured in the same direction as Right Ascension, but is usually measured in degrees. The polar and azimuthal angles carry the somewhat unfortunate names of celestial latitude and celestial longitude respectively in spite of the fact that these names would be more appropriate for Declination and Right Ascension. Again these coordinates may exist in the topocentric, geocentric, heliocentric, or barycentric forms.

c. Alt-Azimuth Coordinate System

The Altitude-Azimuth coordinate system is the most familiar to the general public. The origin of this coordinate system is the observer and it is rarely shifted to any other point. The fundamental plane of the system contains the observer and the horizon. While the horizon is an intuitively obvious concept, a rigorous definition is needed as the apparent horizon is rarely coincident with the location of the true horizon. To define it, one must first define the zenith. This is the point directly over the observer's head, but is more carefully defined as the extension of the local gravity vector outward through the celestial sphere. This point is known as the astronomical zenith. Except for the oblateness of the earth, this zenith is usually close to the extension of the local radius vector from the center of the earth through the observer to the celestial sphere. The presence of large masses nearby (such as a mountain) could cause the local gravity vector to depart even further from the local radius vector. The horizon is then that line on the celestial sphere which is everywhere 90° from the zenith. The altitude of an object is the angular distance of an object above or below the horizon measured along a great circle passing through the object and the zenith. The azimuthal angle of this coordinate system is then just the azimuth of the object. The only problem here arises from the location of the zero point. Many older books on

19

astronomy will tell you that the azimuth is measured westward from the *south* point of the horizon. However, only astronomers did this and most of them don't anymore. Surveyors, pilots and navigators, and virtually anyone concerned with local coordinate systems measures the azimuth from the *north* point of the horizon increasing through the east point around to the west. That is the position that I take throughout this book. Thus the azimuth of the cardinal points of the compass are N(0°), E(90°), S(180°), W(270°).

2.3 Geographic Coordinate Systems

Before leaving the subject of specialized coordinate systems, we should say something about the coordinate systems that measure the surface of the earth. To an excellent approximation the shape of the earth is that of an oblate spheroid. This can cause some problems with the meaning of local vertical.

a. The Astronomical Coordinate System

The traditional coordinate system for locating positions on the surface of the earth is the *latitude-longitude coordinate system*. Most everyone has a feeling for this system as the latitude is simply the angular distance north or south of the equator measured along the local meridian toward the pole while the longitude is the angular distance measured along the equator to the local meridian from some reference meridian. This reference meridian has historically be taken to be that through a specific instrument (the Airy transit) located in Greenwich England. By a convention recently adopted by the International Astronomical Union, longitudes measured east of Greenwich are considered to be positive and those measured to the west are considered to be negative. Such coordinates provide a proper understanding for a perfectly spherical earth. But for an earth that is not exactly spherical, more care needs to be taken.

b. The Geodetic Coordinate System

In an attempt to allow for a nonspherical earth, a coordinate system has been devised that approximates the shape of the earth by an oblate spheroid. Such a figure can be generated by rotating an ellipse about its minor axis, which then forms the axis of the coordinate system. The plane swept out by the major axis of the ellipse is then its equator. This

20

approximation to the actual shape of the earth is really quite
good. The *geodetic latitude* is now given by the angle between
the local vertical and the plane of the equator where the local
vertical is the normal to the oblate spheroid at the point in
question. The *geodetic longitude* is roughly the same as in the
astronomical coordinate system and is the angle between the
local meridian and the meridian at Greenwich. The difference
between the local vertical (i.e., the normal to the local
surface) and the astronomical vertical (defined by the local
gravity vector) is known as the "*deflection of the vertical*"
and is usually less than 20 arcsec. The oblateness of the earth
allows for the introduction of a third coordinate system
sometimes called the *geocentric coordinate system*.

c. The Geocentric Coordinate System

Consider the oblate spheroid that best fits the actual
figure of the earth. Now consider a radius vector from the
center to an arbitrary point on the surface of that spheroid.
In general, that radius vector will not be normal to the
surface of the oblate spheroid (except at the poles and the
equator) so that it will define a different local vertical.
This in turn can be used to define a different latitude from
either the astronomical or geodetic latitude. For the earth,
the maximum difference between the geocentric and geodetic
latitudes occurs at about 45° latitude and amounts to about
11'33". While this may not seem like much, it amounts to about
eleven and a half nautical miles (13.3 miles or 21.4 km.) on
the surface of the earth. Thus, if you really want to know
where you are you must be careful which coordinate system you
are using. Again the *geocentric longitude* is defined in the
same manner as the geodetic longitude, namely it is the angle
between the local meridian and the meridian at Greenwich.

2.4 Coordinate Transformations

A great deal of the practical side of celestial mechanics
involves transforming observational quantities from one
coordinate system to another. Thus it is appropriate that we
discuss the manner in which this is done in general to find the
rules that apply to the problems we will encounter in celestial
mechanics. While within the framework of mathematics it is
possible to define myriads of coordinate transformations, we
shall concern ourselves with a special subset called *linear
transformations*. Such coordinate transformations relate the
coordinates in one frame to those in a second frame by means of
a system of linear algebraic equations. Thus if a vector $\bar{X}$ in

one coordinate system has components X_j, in a primed-coordinate system a vector $\vec{X}'$ to the same point will have components X'_j given by

$$X'_i = \sum_j A_{ij} X_j + B_i \quad . \tag{2.4.1}$$

In vector notation we could write this as

$$\vec{X}' = A\vec{X} + \vec{B} \quad . \tag{2.4.2}$$

This defines the general class of linear transformation where A is some matrix and $\vec{B}$ is a vector. This general linear form may be divided into two constituents, the matrix A and the vector $\vec{B}$. It is clear that the vector $\vec{B}$ may be interpreted as a shift in the origin of the coordinate system, while the elements A_{ij} are the cosines of the angles between the axes X_i and X_j and are called the directions cosines (see Figure 2.1). Indeed, the vector $\vec{B}$ is nothing more than a vector from the origin of the un-primed coordinate frame to the origin of the primed coordinate frame. Now if we consider two points that are fixed in space and a vector connecting them, then the length and orientation of that vector will be independent of the origin of the coordinate frame in which the measurements are made. That places an additional constraint on the types of linear transformations that we may consider. For instance, transformations that scaled each coordinate by a constant amount, while linear, would change the length of the vector as measured in the two coordinate systems. Since we are only using the coordinate system as a convenient way to describe the vector, its length must be independent of the coordinate system. Thus we shall restrict our investigations of linear transformations to those that transform orthogonal coordinate systems while preserving the length of the vector.

Thus the matrix A must satisfy the following condition

$$\vec{X}'.\vec{X}' = (A\vec{X}).(A\vec{X}) = \vec{X}.\vec{X} \quad , \tag{2.4.3}$$

which in component form becomes

$$\sum_i (\sum_j A_{ij} X_i)(\sum_k A_{ik} X_k) = \sum_{jk}(\sum_i A_{ij} A_{ik}) X_j X_k = \sum_i X_i^2 \quad . \tag{2.4.4}$$

This must be true for all vectors in the coordinate system so that

$$\sum_i A_{ij} A_{ik} = \delta_{jk} = \sum_i A_{ji}^{-1} A_{ik} \quad . \tag{2.4.5}$$

Now remember that the Kronecker delta δ_{ij} is the unit matrix

22

and any element of a group that multiplies another and produces that group's unit element is defined as the inverse of that element. Therefore

$$A_{ji} = [A_{ij}]^{-1} \quad .$$

(2.4.6)

Interchanging the elements of a matrix produces a new matrix which we have called the transpose of the matrix. Thus orthogonal transformations that preserve the length of vectors have inverses that are simply the transpose of the original matrix so that

$$A^{-1} = A^{T} \quad .$$

(2.4.7)

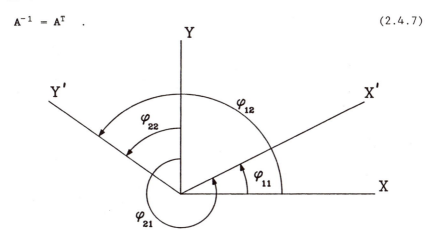

Figure 2.1 shows two coordinate frames related by the transformation angles ϕ_{ij}. Four coordinates are necessary if the frames are not orthogonal.

This means that given that transformation **A** in the linear system of equations (2.4.2), we may invert the transformation, or solve the linear equations, by multiplying those equations by the *transpose* of the original matrix or

$$\vec{X} = A^{T}\vec{X}' - A^{T}\vec{B} \quad .$$

(2.4.8)

Such transformations are called orthogonal unitary transformations, or orthonormal transformations, and the result given in equation (2.4.8) greatly simplifies the process of carrying out a transformation from one coordinate system to another and back again.

We can further divide orthonormal transformations into two categories. These are most easily described by visualizing the relative orientation between the two coordinate systems.

23

Consider a transformation that carries one coordinate into the negative of its counterpart in the new coordinate system while leaving the others unchanged. If the changed coordinate is, say, the x-coordinate, the transformation matrix would be

$$\mathbf{A} = \begin{bmatrix} -1 & 0 & 0 \\ 0 & 1 & 0 \\ 0 & 0 & 1 \end{bmatrix} \quad , \tag{2.4.9}$$

which is equivalent to viewing the first coordinate system in a mirror. Such transformations are known as *reflection transformations* and will take a right handed coordinate system into a left handed coordinate system. The length of any vectors will remain unchanged. The x-component of these vectors will simply be replaced by its negative in the new coordinate system. However, this will not be true of "vectors" that result from the vector cross product. The values of the components of such a vector will remain unchanged implying that a reflection transformation of such a vector will result in the orientation of that vector being changed. If you will, this is the origin of the "right hand rule" for vector cross products. A left hand rule results in a vector pointing in the opposite direction. Thus such vectors are not invariant to reflection transformations because their orientation changes and this is the reason for putting them in a separate class, namely the axial (pseudo) vectors. Since the Levi-Civita tensor generates the vector cross product from the elements of ordinary (polar) vectors, it must share this strange transformation property. Tensors that share this transformation property are, in general, known as tensor densities or pseudo-tensors. Therefore we should call ϵ_{ijk} defined in equation (1.2.7) the Levi-Civita tensor density. Indeed, it is the invariance of tensors, vectors, and scalars to orthonormal transformations that is most correctly used to define the elements of the group called tensors. Finally, it is worth noting that an orthonormal reflection transformation will have a determinant of -1. The unitary magnitude of the determinant is a result of the magnitude of the vector being unchanged by the transformation, while the sign shows that some combination of coordinates has undergone a reflection.

As one might expect, the elements of the second class of orthonormal transformations have determinants of +1. These represent transformations that can be viewed as a rotation of the coordinate system about some axis. Consider a transformation between the two coordinate systems displayed in Figure 2.1. The components of any vector $\vec{C}$ in the primed coordinate system will be given by

$$\begin{pmatrix} C_{x'} \\ C_{y'} \\ C_{z'} \end{pmatrix} = \begin{pmatrix} \cos\phi_{11} & \cos\phi_{12} & 0 \\ \cos\phi_{21} & \cos\phi_{22} & 0 \\ 0 & 0 & 1 \end{pmatrix} \begin{pmatrix} C_x \\ C_y \\ C_z \end{pmatrix} \qquad (2.4.10)$$

If we require the transformation to be orthonormal, then the direction cosines of the transformation will not be linearly independent since the angles between the axes must be $\pi/2$ in both coordinate systems. Thus the angles must be related by

$$\left.\begin{aligned} \phi_{11} &= \phi_{22} \equiv \phi \\ \phi_{12} &= \phi_{11} + \pi/2 = \phi + \pi/2 \\ (2\pi - \phi_{21}) &= \pi/2 - \phi_{22} \Rightarrow \phi_{21} = (\phi + \pi/2) + \pi \end{aligned}\right\} \qquad (2.4.11)$$

Using the addition identities for trigonometric functions, equation (2.4.10) can be given in terms of the single angle ϕ by

$$\begin{pmatrix} C_{x'} \\ C_{y'} \\ C_{z'} \end{pmatrix} = \begin{pmatrix} \cos\phi & \sin\phi & 0 \\ -\sin\phi & \cos\phi & 0 \\ 0 & 0 & 1 \end{pmatrix} \begin{pmatrix} C_x \\ C_y \\ C_z \end{pmatrix} \qquad (2.4.12)$$

This transformation can be viewed as a simple rotation of the coordinate system about the Z-axis through an angle ϕ. Thus,

$$\mathrm{Det}\begin{vmatrix} \cos\phi & \sin\phi & 0 \\ -\sin\phi & \cos\phi & 0 \\ 0 & 0 & 1 \end{vmatrix} = \cos^2\phi + \sin^2\phi = +1 \qquad (2.4.13)$$

In general, the rotation of any cartesian coordinate system about one of its principal axes can be written in terms of a matrix whose elements can be expressed in terms of the rotation angle. Since these transformations are about one of the coordinate axes, the components along that axis remain unchanged. The rotation matrices for each of the three axes are

$$P_x(\phi) = \begin{pmatrix} 1 & 0 & 0 \\ 0 & \cos\phi & \sin\phi \\ 0 & -\sin\phi & \cos\phi \end{pmatrix}$$

$$P_y(\phi) = \begin{bmatrix} \cos\phi & 0 & -\sin\phi \\ 0 & 1 & 0 \\ \sin\phi & 0 & \cos\phi \end{bmatrix}$$

$$P_z(\phi) = \begin{bmatrix} \cos\phi & \sin\phi & 0 \\ -\sin\phi & \cos\phi & 0 \\ 0 & 0 & 1 \end{bmatrix}$$

$$(2.4.14)$$

It is relatively easy to remember the form of these matrices for the row and column of the matrix corresponding to the rotation axis always contains the elements of the unit matrix since that component is not affected by the transformation. The diagonal elements always contain the cosine of the rotation angle while the remaining off diagonal elements always contain the sine of the angle modulo a sign. For rotations about the X- or Z-axes, the sign of the upper right off diagonal element is positive and the other negative. The situation is just reversed for rotations about the Y-axis. So important are these rotation matrices that it is worth remembering their form so that they need not be re-derived every time they are needed.

One can show that it is possible to get from any given orthogonal coordinate system to another through a series of three successive coordinate rotations. Thus a general orthonormal transformation can always be written as the product of three coordinate rotations about the orthogonal axes of the coordinate systems. It is important to remember that the matrix product is not commutative so that the order of the rotations is important. So important is this result, that the angles used for such a series of transformations have a specific name.

2.5 The Eulerian Angles

Leonard Euler proved that the general motion of a rigid body when one point is held fixed corresponds to a series of three rotations about three orthogonal coordinate axes. Unfortunately the definition of the Eulerian angles in the literature is not always the same (see Goldstein[2] p. 108). We shall use the definitions of Goldstein and generally follow them throughout this book. The order of the rotations is as follows. One begins with a rotation about the Z-axis. This is followed by a rotation about the new X-axis. This, in turn, is followed by a rotation about the resulting Z"-axis. The three successive rotation angles are $[\phi,\theta,\psi]$.

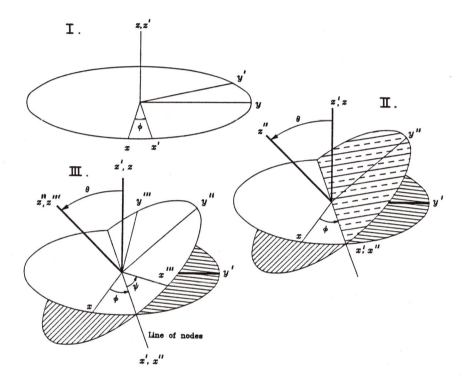

Figure 2.2 shows the three successive rotational transformations corresponding to the three Euler Angles (ϕ, θ, ψ) that can effect a transformation from one orthogonal coordinate frame to another that bears an arbitrary orientation with respect to the first.

This series of rotations is shown in Figure 2.2. Each of these rotational transformations is represented by a transformation matrix of the type given in equation (2.4.14) so that the complete set of Eulerian transformation matrices is

27

$$P_z(\phi) = \begin{pmatrix} \cos\phi & \sin\phi & 0 \\ -\sin\phi & \cos\phi & 0 \\ 0 & 0 & 1 \end{pmatrix}$$

$$P_{x'}(\theta) = \begin{pmatrix} 1 & 0 & 0 \\ 0 & \cos\theta & \sin\theta \\ 0 & -\sin\theta & \cos\theta \end{pmatrix} \qquad (2.5.1)$$

$$P_{z''}(\psi) = \begin{pmatrix} \cos\psi & \sin\psi & 0 \\ -\sin\psi & \cos\psi & 0 \\ 0 & 0 & 1 \end{pmatrix}$$

and the complete single matrix that describes these transformations is

$$A(\phi,\theta,\psi) = P_{z''}(\psi)P_{x'}(\theta)P_z(\phi) \quad . \qquad (2.5.2)$$

Thus the components of any vector $\vec{X}$ can be found in any other coordinate system as the components of $\vec{X}'$ from

$$\vec{X}' = A\vec{X} \quad . \qquad (2.5.3)$$

Since the inverse of orthonormal transformations has such a simple form, the inverse of the operation can easily be found from

$$\vec{X} = A^{-1}\vec{X}' = A^T\vec{X}' = [P_z^T(\phi)P_{x'}^T(\theta)P_{z''}^T(\psi)]\vec{X}' \quad . \qquad (2.5.4)$$

2.6 The Astronomical Triangle

The rotational transformations described in the previous section enable simple and speedy representations of the vector components of one cartesian system in terms of those of another. However, most of the astronomical coordinate systems are spherical coordinate systems where the coordinates are measured in arc lengths and angles. The transformation from one of these coordinate frames to another is less obvious. One of the classical problems in astronomy is relating the defining coordinates of some point in the sky (say representing a star or planet), to the local coordinates of the observer at any given time. This is usually accomplished by means of the Astronomical Triangle which relates one system of coordinates

28

to the other through the use of a spherical triangle. The solution of that triangle is usually quoted *ex cathedra* as resulting from spherical trigonometry. Instead of this approach, we shall show how the result (and many additional results) may be generated from the rotational transformations that have just been described.

Since the celestial sphere rotates about the north celestial pole due to the rotation of the earth, a great circle through the north celestial pole and the object (a meridian) appears to move across the sky with the object. That meridian will make some angle at the pole with the observers local prime meridian (i.e., the great circle through the north celestial pole and the observer's zenith). This angle is known as the local hour angle and may be calculated knowing the object's right ascension and the sidereal time. This latter quantity is obtained from the local time (including date) and the observer's longitude. Thus, given the local time, the observer's location on the surface of the earth (i.e., the latitude and longitude), and the coordinates of the object (i.e., its Right Ascension and declination), two sides and an included angle of the spherical triangle shown in Figure 2.3 may be considered known. The problem then becomes finding the remaining two angles and the included side. This will yield the local azimuth A, the zenith distance z which is the complement of the altitude, and a quantity known as the parallactic angle η. While this latter quantity is not necessary for locating the object approximately in the sky, it is useful for correcting for atmospheric refraction which will cause the image to be slightly displaced along the verticle circle from its true location. This will then enter into the correction for atmospheric extinction and is therefore useful for photometry.

In order to solve this problem, we will solve a separate problem. Consider a cartesian coordinate system with a z-axis pointing along the radius vector from the origin of both astronomical coordinate systems (i.e., equatorial and alt-azimuth) to the point Q. Let the y-axis lie in the meridian plane containing Q and be pointed toward the north celestial pole. The x-axis will then simply be orthogonal to the y- and z-axes. Now consider the components of any vector in this coordinate system. By means of rotational transformations we may calculate the components of that vector in any other coordinate frame. Therefore consider a series of rotational transformations that would carry us through the sides and angles of the astronomical triangle so that we return to exactly the initial xyz coordinate system. Since the series of transformations that accomplish this must exactly reproduce the components of the initial arbitrary vector, the transformation matrix must be the unit matrix with elements δ_{ij}. If we proceed from point Q to the north celestial pole and then on to the

29

zenith, the rotational transformations will involve only quantities related to the given part of our problem [i.e., $(\pi/2-\delta)$, h, $(\pi/2-\phi)$]. Completing the trip from the zenith to Q will involve the three local quantities [i.e., A, $(\pi/2-H)$, η]. The total transformation matrix will then involve six rotational matrices, the first three of which involve given angles and the last three of which involve unknowns and it is this total matrix which is equal to the unit matrix. Since all of the transformation matrices represent orthonormal transformations, their inverse is simply their transpose. Thus we can generate a matrix equation, one side of which involves matrices of known quantities and the other side of which contains matrices of the unknown quantities.

Let us now follow this program and see where it leads. The first rotation of our initial coordinate system will be through the angle $[-(\pi/2-\delta)]$. This will carry us through the complement of the declination and align the z-axis with the rotation axis of the earth. Since the rotation will be about the x-axis, the appropriate rotation matrix from equation (2.4.14) will be

$$
P_x[-(\pi/2-\delta)] = \begin{pmatrix} 1 & 0 & 0 \\ 0 & \sin\delta & -\cos\delta \\ 0 & \cos\delta & \sin\delta \end{pmatrix} \qquad (2.6.1)
$$

Now rotate about the new z-axis that is aligned with the polar axis through a counterclockwise or positive rotation of (h) so that the new y-axis lies in the local prime meridian plane pointing away from the zenith. The rotation matrix for this transformation involves the hour angle so that

$$
P_z[h] = \begin{pmatrix} \cos(h) & \sin(h) & 0 \\ -\sin(h) & \cos(h) & 0 \\ 0 & 0 & 1 \end{pmatrix} \qquad (2.6.2)
$$

Continue the trip by rotating through $[+(\pi/2-\phi)]$ so that the z-axis of the coordinate system aligns with a radius vector through the zenith. This will require a positive rotation about the x-axis so that the appropriate transformation matrix is

$$
P_x[+(\pi/2-\phi)] = \begin{pmatrix} 1 & 0 & 0 \\ 0 & \sin\phi & \cos\phi \\ 0 & -\cos\phi & \sin\phi \end{pmatrix} \qquad (2.6.3)
$$

30

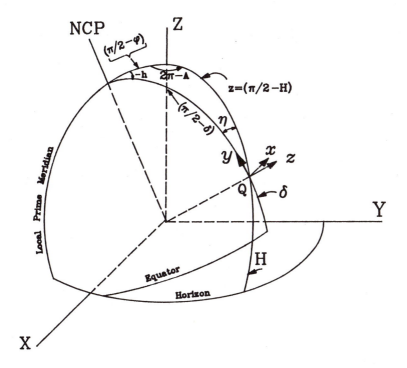

Figure 2.3 shows the Astronomical Triangle with the zenith in the Z-direction. The solution of this triangle is necessary for transformations between the Alt-Azimuth coordinate system and the Right Ascension-Declination coordinate system. The latter coordinates are found from the hour angle h and the distance from the North Celestial Pole.

Now rotate about the z-axis through the azimuth $[2\pi\text{-}A]$ so that the y-axis will now be directed toward the point in question Q. This is another z-rotation so that the appropriate transformation matrix is

$$\mathbf{P_z}\,[2\pi\text{-}A] = \begin{bmatrix} \cos(A) & -\sin(A) & 0 \\ \sin(A) & \cos(A) & 0 \\ 0 & 0 & 1 \end{bmatrix} \qquad (2.6.4)$$

31

We may return the z-axis to the original z-axis by a negative rotation about the x-axis through the zenith distance $(\pi/2-H)$ which yields a transformation matrix

$$P_x[-(\pi/2-H)] = \begin{pmatrix} 1 & 0 & 0 \\ 0 & \sin(H) & -\cos(H) \\ 0 & \cos(H) & \sin(H) \end{pmatrix} \quad (2.6.5)$$

Finally the coordinate frame may be aligned with the starting frame by a rotation about the z-axis through an angle $[\pi+\eta]$ yielding the final transformation matrix

$$P_z[+(\pi+\eta)] = \begin{pmatrix} -\cos\eta & -\sin\eta & 0 \\ +\sin\eta & -\cos\eta & 0 \\ 0 & 0 & 1 \end{pmatrix} \quad (2.6.6)$$

Since the end result of all these transformations is to return to the starting coordinate frame, the product of all the transformations yields the identity matrix or

$$P_z[+(\pi+\eta)]P_x[-z]P_z[2\pi-A]P_x[+(\pi/2-\phi)]P_z[h]P_x[\delta-\pi/2] = 1. \quad (2.6.7)$$

We may separate the knowns from the unknowns by remembering that the inverse of an orthonormal transformation matrix is its transpose so that

$$P_z[+(\pi+\eta)]P_x[H-\pi/2]P_z[2\pi-A] = P_x^T[\delta-\pi/2]P_z^T[h]P_x^T[\pi/2-\phi]. \quad (2.6.8)$$

We must now explicitly perform the matrix products implied by equation (2.6.8) and the nine elements of the left hand side must equal the nine elements of the right hand side. These nine relations provide in a natural way all of the relations possible for the spherical triangle. These, of course, include the usual relations quoted for the solution to the astronomical triangle. These nine relations are

$$\sin(H) = [\cos(h)\cos\phi\cos\delta + \sin\delta\sin\phi]$$

$$\cos(H)\cos(A) = [\sin\delta\cos\phi - \cos(h)\cos\delta\sin\phi]$$

$$\cos(H)\sin(A) = -\cos\delta\sin(h)$$

$$\cos(H)\cos\eta = \cos(h)\sin\delta\cos\phi - \cos\delta\sin\phi$$

$$\cos(H)\sin\eta = \sin(h)\cos\phi$$

$$[\sin(A)\sin\eta + \sin(H)\cos(A)\cos\eta] =$$
$$- [\cos\delta\cos\phi + \cos(h)\sin\phi\sin\delta]$$

$$[\sin(A)\cos\eta - \sin(H)\cos(A)\sin\eta] = -\sin\phi\sin(h)$$

$$[\cos(A)\sin\eta - \sin(H)\sin(A)\cos\eta] = +\sin\delta\sin(h)$$

$$[\cos(A)\cos\eta + \sin(H)\sin(A)\sin\eta] = -\cos(h)$$

(2.6.9)

Since the altitude is defined to lie in the first or fourth quadrants, the first of these relations uniquely specifies H. The next two will then uniquely give the azimuth A and the following two allow for the unique specification of the parallactic angle. Thus these relations are sufficient to effect the coordinate transformation from either the defining coordinate frame to the observer's frame or vice versa. However, the more traditional solution of the astronomical triangle can be found from

$$P_z[+(\pi+\eta)]P_x[H-\pi/2]P_z[2\pi-A]P_x[\pi/2-\phi] = P_x^T[\delta-\pi/2]P_z^T[h], \quad ,(2.6.10)$$

where only the last row of the matrices is considered. These elements yield

$$\sin(h)\cos\delta = - \sin(A)\cos(H)$$

$$\cos(h)\cos\delta = - \cos(A)\cos(H)\sin\phi + \sin(H)\cos\phi$$

$$\sin\delta = +\cos(A)\cos(H)\cos\phi +\sin(H)\sin\phi$$

(2.6.11)

These results differ from those found in some astronomical textbooks as we have defined the azimuth from the north point. So to get the traditional results we would have to replace A by $[\pi-A]$. Having discussed how we locate objects in the sky in different coordinate frames and how to relate those frames, we will now turn to a brief discussion of how they are to be located in time.

2.7 Time

The independent variable of Newtonian mechanics is time and thus far we have said little about it. Newton viewed time as absolute and 'flowing' uniformly throughout all space. This intuitively reasonable view was shown to be incorrect in 1905 by Albert Einstein in the development of what has become known as the Special Theory of Relativity. However, the problems introduced by special relativity are generally small for objects moving in the solar system. What does complicate the concept of time is the less sophisticated notion of how it is measured. As with other tangled definitions of science, historical developments have served to complicate immensely the definition of what ought to be a simple concept. We will choose to call the units of time seconds, minutes, hours, days, years, and centuries (there are others, but we will ignore them for this book). The relationships between these units are not simple and have been dictated by history.

In some very broad sense, time can be defined in terms of an interval between two events. The difficulty arises when one tries to decide what events should be chosen for all to use. In other words, what "clock" shall we use to define time? Clocks run in response to physical forces so we are stuck with an engineering problem of finding the most accurate clock. Currently, the most accurate clocks are those that measure the interval between atomic processes and have an accuracy of the order of 1 part in 10^{11} to 1 part in 10^{15}. Clocks such as these form the basis for measuring time and time kept by them is known as international atomic time (TAI for short). However, the world for centuries has kept time by clocks that mimic the rising and setting of the sun or rotation of the earth. Certainly prehistoric man realized that all days were not of equal length and therefore could not serve to define a unit of time. However, the interval between two successive transits (crossings of the local meridian) of the sun is a more nearly constant interval. If the orbit of the earth were perfectly circular, then the motion of the sun along the ecliptic would be uniform in time. Therefore, it could not also be uniform along the equator. This nonuniformity of motion along the equator will lead to differences in successive transits of the sun. To make matters worse, the orbit of the earth is elliptical so that the motion along the ecliptic is not even uniform. One could correct for this or choose to keep time by the stars.

Time that is tied to the apparent motion of the stars is called sidereal time and the local sidereal time is of importance to astronomers as it defines the location of the origin of the Right Ascension - Declination coordinate frame as seen by a local observer. It therefore determines where things are in the sky. Local sidereal time is basically defined as the hour angle of the vernal equinox as seen by the observer.

However, as our ability to measure intervals of time became more precise, it became clear that the earth did not rotate at a constant rate. While a spinning object would seem to provide a perfect clock as it appears to be independent of all of the forces of nature, other objects acting through those forces cause irregularities in the spin rate. In fact, the earth makes a lousy clock. Not only does the rotation rate vary, but the location of the intersection of the north polar axis with the surface of the earth changes by small amounts during the year. In addition long term precession resulting from torques generated by the sun and moon acting on the equatorial bulge of the earth, cause the polar axis, and hence the vernal equinox, to change its location among the stars. This, in turn, will influence the interval of time between successive transits of any given star. Time scales based on the rotation of the earth do not correspond to the uniformly running time envisioned by Newton. Thus we have need for another type of time, a dynamical time suitable for expressing the solution to the Newtonian equations of motion for objects in the solar system. Such time is called terrestrial dynamical time (TDT) and is an extension of what was once known as ephemeris time (ET), abandoned in 1984. Since it is to be the smoothly flowing time of Newton, it can be related directly to atomic time (TAI) with an additive constant to bring about agreement with the historical ephemeris time of 1984. Thus we have

$$TDT \equiv TAI + 32.184 \text{ seconds.} \qquad (2.7.1)$$

Unfortunately, we and the atomic clocks are located on a moving body with a gravitational field and both these properties will affect the rate at which clocks run compared with similar clocks located in an inertial frame free of the influence of gravity and accelerative motion. Thus to define a time that is appropriate for the navigation of spacecraft in the solar system, we must correct for the effects of special and general relativity and find an inertial coordinate frame in which to keep track of the time. The origin of such a system can be taken to be the barycenter (center of mass) of the solar system and we can define barycentric dynamical Time (TDB) to be that time. The relativistic terms are small indeed so that the difference between TDT and TDB is less than .002 sec. A specific formula for calculating it is given in the Astronomical Almanac[3] . Essentially, terrestrial dynamical time is the time used to calculate the motion of objects in the solar system. However, it is only approximately correct for observers who would locate objects in the sky. For this we need another time scale that accounts for the irregular rotation of the earth.

Historically, such time was known as Greenwich mean time, but this term has been supplanted by the more grandiose sounding universal time (UT). The fundamental form of universal time (UT1) is used to determine the civil time standards and is determined from the transits of stars. Thus it is related to Greenwich mean sidereal time and contains nonuniformities due to variations in the rotation rate of the earth. This is what is needed to find an object in the sky. Differences between universal time and terrestrial dynamical time are given in the Astronomical Almanac and currently (1988) amount to nearly one full minute as the earth is "running slow". Of course determination of the difference between the dynamical time of theory and the observed time dictated by the rotation of the earth must be made after the fact, but the past behavior is used to estimate the present.

Finally, the time that serves as the world time standard and is broadcast by WWV and other radio stations is called coordinated universal time (UTC) and is arranged so that

$$|UT1-UTC| < 1 \text{ second} \quad . \tag{2.7.2}$$

Coordinated universal time flows at essentially the rate of atomic time (give or take the relativistic corrections), but is adjusted by an integral number of seconds so that it remains close to UT1. This adjustment could take place as often as twice a year (on December 31 and June 30) and results in a systematic difference between UTC and TAI. This difference amounted to 10 sec. in 1972. From then to the present (1988), corrections amounting to an additional 14 sec. have had to be made to maintain approximate agreement between the heavens and the earth.

Coordinated universal time is close enough to UT1 to locate objects in the sky and its conversion to local sidereal time in place of UT1 can be effectively arrived at by scaling by the ratio of the sidereal to solar day. Due to a recent adoption by the International Astronomical Union that terrestrial Longitude will be defined as increasing positively to the *east*, local mean solar time will just be

$$(LMST) = (UTC) + \lambda \quad , \tag{2.7.3}$$

where λ is the longitude of the observer. The same will hold true for sidereal time so that

$$LST = GST + \lambda \quad , \tag{2.7.4}$$

where the Greenwich sidereal time (GST) can be obtained from

36

UT1 and the date. The local sidereal time is just the local hour angle of the vernal equinox by definition so that the hour angle of an object is

$$h = LST - R.A. \tag{2.7.5}$$

Here we have taken hour angles measured west of the prime meridian as increasing.

Given the appropriate time scale, we can measure the motion of objects in the solar system (TDT) and find their location in the sky (UT1 and LST). There are numerous additional small corrections including the barycentric motion of the earth (its motion about the center of mass of the earth-moon system) and so forth. For those interested in time to better than a millisecond all of these corrections are important and constitute a study in and of themselves. For the simple acquisition of celestial objects in a telescope, knowledge of the local sidereal time as determined from UTC will generally suffice.

Chapter 2: Exercises

1. Transform from the Right Ascension-Declination coordinate system (α, δ) to Ecliptic coordinates (λ, β) by a rotation matrix. Show all angles required and give the transformation explicitly.

2. Given two n-dimensional coordinate systems X_i and X_j and an orthonormal transformation between the two A_{ij}, prove that exactly $n(n-1)/2$ terms are required to completely specify the transformation.

3. Find the transformation matrix appropriate for a transformation from the Right Ascension-Declination coordinate system to the Alt-Azimuth coordinate system.

4. Consider a space shuttle experiment in which the pilot is required to orient the spacecraft so that its major axis is pointing at a particular point in the sky (i.e. α, δ). Unfortunately his yaw thrusters have failed and he can only roll and pitch the spacecraft. Given that the spacecraft has an initial orientation $(\alpha_1, \delta_1,$ and α_2, δ_2 of the long and short axes of the spacecraft) and must first roll and then pitch to achieve the desired orientation, find the roll and pitch angles (ξ, η) that the pilot must move the craft through in order to carry out the maneuver.

3

The Basics of Classical Mechanics

Celestial mechanics is a specialized branch of classical mechanics and a proper understanding of the subject requires that one see how it is embedded in this larger subject. One might describe the fundamental problem of celestial mechanics as the description of the motion of celestial objects that move under the influence of the gravitational forces present in the solar system. The approach of classical mechanics to this problem would be to divide it into two parts. The first of these would be to write the equations of motion for an object moving under the influence of an arbitrary collection of mass points. The second part then consists of solving those equations. Therefore it is appropriate that we spend some time with the fundamentals of classical mechanics so that their relation to the more specific subject of celestial mechanics is clear. The most basic concept of all of theoretical physics is the notion of a conservation law. But before one can discuss the conservation of quantities, one must define them.

3.1 Newton's Laws and the Conservation of Momentum and Energy

Newton's famous laws of motion can largely be taken as a set of definitions. Consider the second law

$$\vec{F} = m\vec{a} = \frac{d(m\vec{v})}{dt} \equiv \frac{d\vec{p}}{dt} \quad . \tag{3.1.1}$$

Here $\vec{p}$ is defined as the linear momentum. It is a simple matter to describe operationally what is meant by mass m and

39

acceleration $\vec{a}$. However, a clear operational definition of what is meant by force that does not make use of some form of equation (3.1.1) is more difficult. However, one can use the first law to describe a situation wherein forces are absent, and that is sufficient to arrive at a conservation law for linear momentum, namely

$$\left.\begin{array}{l} \vec{F} = 0 = d\vec{p}/dt \\[10pt] \vec{p} = \text{const.} \end{array}\right\} \qquad . \tag{3.1.2}$$

We may use the concept of linear momentum to define an additional quantity called *angular momentum* $(\vec{L})$ so that

$$\vec{L} \equiv \vec{r} \times \vec{p} \qquad . \tag{3.1.3}$$

Since $\vec{p}$ is a conserved quantity, it would seem plausible that angular momentum will also be a conserved quantity.

Let us define a force-like quantity, by analogy with the angular momentum, called the *torque* as

$$\vec{N} \equiv \vec{r} \times \vec{F} = \vec{r} \times d\vec{p}/dt = \frac{d}{dt}(\vec{r} \times \vec{p}) - \left(\frac{d\vec{r}}{dt} \times \vec{p}\right) \qquad . \tag{3.1.4}$$

However,

$$\frac{d\vec{r}}{dt} \times \vec{p} = \vec{v} \times m\vec{v} = 0 \qquad . \tag{3.1.5}$$

Therefore

$$\vec{N} = d\vec{L}/dt \qquad . \tag{3.1.6}$$

Thus a force free situation will also be a torque free situation and the angular momentum will be constant and conserved in exactly the same sense as the linear momentum.

Finally, let us define the concept of *work* as the line-integral of force over some path or

$$W_{a,b} \equiv \int_a^b \vec{F} \cdot d\vec{s} \qquad . \tag{3.1.7}$$

Note that the quantity called work is a scalar quantity as only the component of the force directed along the path contributes to the work. We may use the notion of work to say something about the nature of the forces along the path. Specifically, if

40

no net work is done while completing a closed path so that

$$\oint \vec{F}.\vec{ds} = 0 \quad , \qquad\qquad (3.1.8)$$

and this is true for *any closed path*, the force is said to be a conservative force. Now there is a theorem in mathematics known as Stokes theorem where

$$\int_c \vec{Q}.\vec{ds} = \int_s (\nabla \times \vec{Q}).\vec{dA} \quad . \qquad\qquad (3.1.9)$$

The left hand quantity is a line integral along some curve C which bounds a surface S. The quantity on the right hand side is a surface integral over that surface where $\vec{dA}$ is a unit vector normal to the differential area dA. Applying this theorem to equation (3.1.8) we get

$$\oint_c \vec{F}.\vec{ds} = \oint_s (\nabla \times \vec{F}).\vec{dA} \quad . \qquad\qquad (3.1.10)$$

But for conservative force fields this result must be true for all paths and hence the right hand side must hold for all enclosed areas. This can only be true if the integrand of the right hand integral is itself zero so that

$$\nabla \times \vec{F} = 0 \quad . \qquad\qquad (3.1.11)$$

Since the curl of the gradient is always zero (i.e., ∇ points in the same direction as itself), we can write a conservative force as the gradient of some scalar V so that

$$\vec{F} = -\nabla V \quad . \qquad\qquad (3.1.12)$$

The quantity V is called the potential energy. Since the quantity $\vec{F}$ is related to operationally defined parameters, the potential energy is defined only through its gradient. Thus we may add or subtract any constant to the potential energy without affecting measurable quantities. This is often done for convenience and the additive constant must be included in any self consistent use of the potential. From equations (3.1.7) and (3.1.12) it is clear that we can determine the work done in moving from point a to point b in terms of the potential as

$$W_{a,b} = \int_a^b \vec{F} \cdot \vec{ds} = -\int_a^b \nabla V \cdot \vec{ds} = -\int_a^b \sum_i \frac{\partial V}{\partial x_i} \hat{x}_i \cdot \vec{ds} \quad . \tag{3.1.13}$$

However,

$$\vec{ds} \equiv \sum_j \hat{x}_j \, dx_j \quad , \tag{3.1.14}$$

so that

$$W_{a,b} = -\int_a^b dV = V(a) - V(b) \quad . \tag{3.1.15}$$

Thus the amount of work done on an object is simply equal to the change in the potential energy in going from a to b.
    Now we may write

$$\int_a^b \vec{F} \cdot \vec{ds} = \int_a^b \frac{d\vec{p}}{dt} \cdot \vec{ds} = \int_a^b m \frac{d\vec{v}}{dt} \cdot \vec{v} dt = \tfrac{1}{2} m \int_a^b v^2 \, dt \quad , \tag{3.1.16}$$

so that the change in the kinetic energy of the particle in going from a to b is

$$T(b) - T(a) = V(a) - V(b) \quad . \tag{3.1.17}$$

Thus the sum of the kinetic and potential energies E is the same at points a and b so that

$$T(b) + V(b) = T(a) + V(a) \quad . \tag{3.1.18}$$

This is nothing more than a statement of the conservation of energy. Clearly energy conservation is a weaker conservation law than conservation of momentum as we had to assume that the force field was conservative in order to obtain it.

3.2 Virtual Work, D'Alembert's Principle, and Lagrange's Equations of Motion

    Consider a system of particles that are not subject to any constraints. A constraint is something that cannot be represented *in a general way* by the forces acting on all the objects. For example, an object moving under the influence of gravity but constrained to roll on the surface of a sphere for part of its motion would be said to be subject to constraints imposed by the sphere. Such constraints imply that forces are

42

acting on the object in such a way as to constrain the motion, but the forces are not known *a priori* and must be found as part of the problem. Only their effect on the object (i.e., its constrained motion) is known. Some constraints can be expressed in terms of the coordinates of the problem and are known as *holonomic constraints*. Constraints that cannot be written in terms of the coordinates alone are called *nonholonomic constraints*. The rolling motion of an object where there is no slippage is an example. The constraint here is on the velocity of the point in contact with the surface. We will leave the consideration of such systems for an advanced mechanics course.

The notion of virtual work is a creation of James Bernoulli and comes about from considering infinitesimal displacements of the particles that are subject to forces $\vec{F}_i$. We can call these displacements $\delta\vec{r}_i$. If these infinitesimal displacements can be called virtual displacements, then $\vec{F}_i.\delta\vec{r}_i$ can be called the virtual work done on the ith particle. The virtual nature of these displacements becomes clear when we require that the forces F_i do not change in response to the virtual displacements δr_i in contrast to the case for real displacements. For a system in equilibrium, the forces on the individual particles vanish and therefore so does the virtual work. For a dynamical system subject to Newton's laws of motion we can say that the forces are balanced by the accelerative response of the system so that

$$\sum_i (\vec{F}_i - \dot{\vec{p}}_i).\delta\vec{r}_i = 0 \quad .$$

$$(3.2.1)$$

This is known as D'Alembert's principle and is useful for what we can derive from it.

It may be the case that the $\vec{r}_i$'s are not all linearly independent. To this point, the choice of the coordinate system used to represent the motion of the system has been arbitrary. Thus it is entirely possible that the coordinates of choice will not be independent of one another. Holonomic constraints can also produce a set of coordinates that are not linearly independent. However, we can hardly expect to unravel the dynamical motion of a system of particles if the coordinates chosen to represent them depend on each other. Therefore, we shall require that the coordinates chosen to represent the system are indeed linearly independent. We shall call any set of coordinates that are linearly independent and describe all the particles of the system *generalized coordinates*. Now consider a transformation from our initial arbitrary set of coordinates to a set of coordinates which are linearly independent and which we shall denote by q_i. We could write such a transformation as

$$\delta \vec{r}_i = \sum_j \frac{\partial \vec{r}_i}{\partial q_j} dq_j \quad . \tag{3.2.2}$$

However, since the q_j's are linearly independent we get

$$\delta q_j = \sum_k \frac{\partial q_j}{\partial q_k} \delta q_k = \sum_k \delta_{jk} dq_k = dq_j \quad . \tag{3.2.3}$$

Substitution of this into D'Alembert's principle gives

$$\sum_j \left[\sum_i \vec{F} \cdot \frac{\partial \vec{r}_i}{\partial q_j} \right] \delta q_i - \sum_i m_i \frac{d^2 \vec{r}_i}{dt^2} \sum_j \frac{\partial \vec{r}_i}{\partial q_j} \delta q_i =$$

$$\sum_j Q_j \delta q_j - \sum_j \sum_i m_i \frac{d}{dt} \left(\frac{d\vec{r}_i}{dt} \cdot \frac{\partial \vec{r}_i}{\partial q_j} - \frac{d\vec{r}_i}{dt} \cdot \frac{d}{dt} \left[\frac{\partial v_i}{\partial q_j} \right] \right) \delta q_j \quad (3.2.4)$$

Here we have used Q_j to stand for the term in parentheses on the left hand side of the equation. Now since the velocity of any particle can be written in terms of the generalized coordinates as

$$\vec{v}_i = \sum_j \frac{\partial \vec{r}_i}{\partial q_j} \frac{dq_j}{dt} + \frac{\partial \vec{r}}{\partial t} \tag{3.2.5}$$

we may calculate its partial derivative with respect to the *total* time derivative of the generalized coordinates as

$$\partial \vec{v}_i / \partial \dot{q}_k = \sum_j [\partial \vec{r}_i / \partial q_j][\partial / \partial \dot{q}_k][dq_j/dt]$$
$$= \sum_j [\partial \vec{r}_i / \partial \dot{q}_j][\partial q_j / \partial q_k] = \partial \vec{r}_i / \partial q_k . \tag{3.2.6}$$

This allows us to write the expanded form of D'Alembert's principle as

$$\sum_i (\vec{F_i} - \dot{\vec{p}}_i) \cdot \delta \vec{r_i} = 0 = \sum_i \left\{ \sum_j Q_j \delta q_j - \sum_j \left[\frac{d}{dt} [m_i \vec{v_i} \cdot (\partial \vec{v_i} / \partial \dot{q_i})] \right. \right.$$

$$\left. \left. - \left[m_i \vec{v_i} \cdot \frac{\partial \vec{v_i}}{\partial q_j} \right] \right] \delta q_j \right\} \qquad (3.2.7)$$

From the definition of kinetic energy, we get

$$\sum_j \left[\frac{d}{dt} [\partial T / \partial q_j] - \frac{\partial T}{\partial q_j} - Q_j \right] \delta q_j = 0 \qquad (3.2.8)$$

However, this result must be true for arbitrary virtual displacements of the generalized coordinates δq_j. Such can only be the case if it is true for each term of the sum. Therefore

$$\frac{d}{dt} [\partial T / \partial \dot{q_j}] - [\partial T / \partial q_j] - Q_j = 0 \qquad (3.2.9)$$

From the definition of Q_j [see equation (3.2.4)] we can write for conservative forces that

$$Q_j = \sum_i \vec{F} \cdot [\partial \vec{r_i} / \partial q_i] = -\sum_i [\nabla V] \cdot [\partial \vec{r_i} / \partial q_j]$$

$$= -\sum_i [\partial V / \partial r_i][\partial r_i / \partial q_i] = -\partial V / \partial q_j . \qquad (3.2.10)$$

Very rarely is the potential energy an *explicit* function of time so that

$$\partial V / \partial \dot{q_j} = 0 . \qquad (3.2.11)$$

From equations (3.2.10) and equation (3.2.11) it is clear that we can combine the potential energy with the kinetic energy in equation (3.2.9) and thereby eliminate the Q_js. So define

$$\mathcal{L} \equiv T - V , \qquad (3.2.12)$$

which is known as the Lagrangian. In terms of the Lagrangian, equation (3.2.9) becomes

$$\frac{d}{dt} [\partial \mathcal{L} / \partial \dot{q_j}] - [\partial \mathcal{L} / \partial q_j] = 0 . \qquad (3.2.13)$$

45

These are known as Lagrange's equations of motion and their solution constitutes the solution of the first part of the basic problem of classical mechanics. The utility of the Lagrangian equations of motion is clear. Given a set of coordinates that are linearly independent, find an expression for the kinetic and potential energies in terms of those coordinates and their time derivatives. Equation (3.2.13) then provides a mechanical means for generating the equations of motion for the particle of interest in the chosen coordinates.

Let us consider as an example the equations of motion for two point masses moving under the influence of their mutual self-gravity. For a generalized set of coordinates, let us use cartesian coordinates with the origin at the center of mass of the system. Further let the mass of one particle be m and the total mass of the system be M. Since this is an isolated system, the motion of the center of mass can be taken to be zero. We can always transform to an inertial frame that moves with the center of mass. Thus the kinetic and potential energies can be written as

$$
\left.
\begin{aligned}
T &= \tfrac{1}{2}mv^2 + \tfrac{1}{2}M(0)^2 = \tfrac{1}{2}\Sigma_j m(dx_j/dt)^2 \\
V &= -\, GmM/[\Sigma_j (x_j^2)^{\frac{1}{2}}]
\end{aligned}
\right\}
\qquad (3.2.14)
$$

and the Lagrangian becomes

$$
\mathcal{L} = \tfrac{1}{2}\Sigma_j m(dx_j/dt)^2 + GmM/(\Sigma_j x_j^2)^{\frac{1}{2}}
\qquad (3.2.15)
$$

Substituting this into equation (3.2.13) and remembering that q_i is x_i we can write the equations of motion as

$$
m\frac{d^2 x_i}{dt^2} + \frac{GmM x_i}{[\Sigma_j x_j^2]^{3/2}} = 0
\qquad (3.2.16)
$$

Before turning to the problem of determining the potential for an arbitrary collection of mass points we will briefly discuss a related method of obtaining the equations of motion.

3.3 The Hamiltonian

There is an approach to mechanics due to Sir W. R. Hamilton that is very similar to Lagrange's and has wide ranging applications in theoretical physics. The Hamiltonian formulation adds nothing new in the form of physical laws, but provides what many feel is a much more powerful formalism with which any student of the physical sciences should be familiar.

The basic idea of the Hamiltonian formulation is to write equations of motion in terms of coordinates and the momentum instead of the coordinates and their time derivatives. Lagrange's equations of motion are second order differential equations requiring 6N constants of integration, which are usually the initial values of $\dot{q}_i$ and q_i. If we choose as generalized coordinates of the problem (q_i, p_i, t), then we can in principle write 2N first order equations of the motion still requiring 6N constants, but reducing the order of the equations to be solved. This can be accomplished by subjecting the Lagrangian equations of motion to a transformation known as the Legendre transformation. First, let us *define* what we will mean by the generalized momenta as

$$p_i = \partial \mathcal{L}/\partial \dot{q}_i \quad . \tag{3.3.1}$$

Lagrange's equations and this definition allow us to write

$$\dot{p}_i = \partial \mathcal{L}/\partial q_i \quad . \tag{3.3.2}$$

Now let us just guess a function of the form

$$H(p,q,t) = \sum_i \dot{q}_i p_i - \mathcal{L}(q,\dot{q},t) \quad , \tag{3.3.3}$$

and call it the Hamiltonian. The total differential of the right hand side of equation (3.3.3) yields

$$dH = \sum_i \dot{q}_i dp_i + \sum_i p_i d\dot{q}_i - \sum_i [\partial \mathcal{L}/\partial \dot{q}_i] d\dot{q}_i - \sum_i [\partial \mathcal{L}/\partial q_i] dq_i$$
$$- [\partial \mathcal{L}/\partial t] dt \quad . \tag{3.3.4}$$

Now the second and third terms cancel by virtue of the definition of the generalized momenta [equation (3.3.1) multiplied by $d\dot{q}_i$ and summed over i]. The partial derivative of the Lagrangian in the fourth term can be replaced by equation (3.3.2) so that the total differential of H becomes

$$dH = \sum_i \dot{q}_i dp_i - \sum_i \dot{p}_i dq_i - [\partial \mathcal{L}/dt] dt = \sum_i [\partial H/\partial q_i] dq_i + \sum_i [\partial H/\partial p_i] dp_i$$
$$+ [\partial H/\partial t] dt \quad . \tag{3.3.5}$$

The right hand part of equation (3.3.5) is just the definition of a total differential. Since the generalized coordinates are linearly independent, the three terms on each side of equation (3.3.5) must be separately equal so that

$$\left. \begin{array}{l} \dot{q}_i = \partial H/\partial p_i \\[2mm] \dot{p}_i = -\partial H/\partial q_i \\[2mm] \partial \mathcal{L}/\partial t = -\partial H/\partial t \end{array} \right\}$$ (3.3.6)

These equations are known as the *canonical equations of Hamilton* and they form a set of 2n first order equations for the motion of the constituents of the system.

From the definition of a total time derivative we can write

$$\frac{dH}{dt} = \sum \left[\frac{\partial H}{\partial q_i} \dot{q}_i + \frac{\partial H}{\partial p_i} \dot{p}_i \right] + \frac{\partial H}{\partial t} \quad .$$ (3.3.7)

However, by substituting for $\dot{q}_i$ and $\dot{p}_i$ from the Hamilton equations of motion, we see that the term in parentheses vanishes so that

$$\frac{dH}{dt} = \frac{\partial H}{\partial t} = -\frac{\partial \mathcal{L}}{\partial t}$$ (3.3.8)

Thus if the Lagrangian is not an *explicit* function of time, then the Hamiltonian will not vary with time at all and will therefore be a constant of the motion. If, in addition, the transformation to the generalized coordinates also does not depend explicitly on time, the Hamiltonian will be the total energy of the system. In most celestial mechanics problems this is indeed the case. The potential depends only on position and not explicitly on time and the generalized coordinates are usually the position coordinates themselves. Thus the Hamiltonian is a constant of the system and is equal to the total energy. The primary exception to this is when analysis is done in a rotating or noninertial coordinate frame. Then the transformation to the generalized coordinates does explicitly involve time and thus the Hamiltonian is not the total energy of the system. However, if the Lagrangian is not an explicit function of time, the Hamiltonian is still a constant of the motion.

One standard way of proceeding with a classical mechanics problem is to find the Lagrangian by determining the potential. Then equation (3.3.3) can be used directly to calculate the Hamiltonian. Then equations (3.3.6) yield the equations of motion and usually one of the constants of the motion has been found in the process. This formalism is so powerful that it forms the basis for a great deal of quantum mechanics. It is

48

clear that for celestial mechanics, the central remaining problem to finding the equations of motion is the determination of the potential and so in the next chapter we will turn to various methods by which that can be done.

Chapter 3: Exercises

1. The escape velocity from the Earth is the minimum
    velocity required to escape the influence of the Earth's
    gravitational field. Neglecting atmospheric drag, use
    basic conservation laws to find the value for the escape
    velocity from the surface of the Earth.

2. a: Find the equations of motion for a rocket projected
       vertically from the surface of the Earth. Again,
       neglect atmospheric drag.

    b: Assuming the rate of mass loss from the rocket is
       constant and equal to 1/60 of the initial mass per
       sec, show that if the exhaust velocity is 2073 m/s,
       then for the rocket to reach escape velocity, the
       ratio of the mass of the fuel to the empty rocket must
       be about 300.

3. Consider a system of n particles moving under the
    influence of gravity alone.

    a: Write down the Lagrangian for the system.
    b: Find the equations of motion in cartesian coordinates
       for the system .

4. Consider a single particle moving under the influence of
    the potential Φ where

    $$\Phi = \sum_{k=0}^{n} \frac{a_k \cos(k\theta)}{(k+1)} \quad .$$

    Find the Lagrangian, Hamiltonian, and the equations of
    motion in spherical coordinates for the particle.

4

Potential Theory

We have seen how the solution of any classical mechanics problem is first one of determining the equations of motion. These then must be solved in order to find the motion of the particles that comprise the mechanical system. In the previous chapter, we developed the formalisms of Lagrange and Hamilton which enable the equations of motion to be written down as either a set of n second order differential equations or 2n first order differential equations depending on whether one chooses the formalism of Lagrange or Hamilton. However, in the methods developed, the Hamiltonian required knowledge of the Lagrangian, and the correct formulation of the Lagrangian required knowledge of the potential through which the system of particles move. Thus, the development of the equations of motion has been reduced to the determination of the potential; the rest is manipulation. In this way the more complicated vector equations of motion can be obtained from the far simpler concept of the scalar field of the potential.

To complete this development we shall see how the potential resulting from the sources of the forces that drive the system can be determined. In keeping with the celestial mechanics theme we shall restrict ourselves to the forces of gravitation although much of the formalism had its origins in the theory of electromagnetism - specifically electrostatics. The most notable difference between gravitation and electromagnetism (other than the obvious difference in the strength of the force) is that the sources of the gravitational force all have the same sign, but all masses behave as if they were attractive.

4.1 The Scalar Potential Field and the Gravitational Field

In the last chapter we saw that any forces with zero curl could be derived from a potential so that if

$$\nabla \times \vec{F} = 0 \quad , \tag{4.1.1}$$

then

$$\vec{F} = - \nabla V \quad , \tag{4.1.2}$$

where V is the potential energy. Forces that satisfy this condition were said to be conservative so that the total energy of the system was constant. Such is the case with the gravitational force. Let us define the gravitational potential energy as Ω so that the gravitational force will be

$$\vec{F} = - \nabla \Omega \quad . \tag{4.1.3}$$

Now by analogy with the electromagnetic force, let us define the gravitational field $\vec{G}$ as the gravitational force per unit mass so that

$$\vec{G} = \vec{F}/m = -(\nabla \Omega)/m \equiv \nabla \Phi \quad . \tag{4.1.4}$$

Here Φ is known as the gravitational potential, and from the form of equation (4.1.4) we can draw a direct comparison to electrostatics. The $\vec{G}$ is analogous to the electric field while Φ is analogous to the electric potential.

Now Newtonian gravity says that the gravitational force between any two objects is proportional to the product of their masses and inversely proportional to the square of the distance separating them and acts along the line joining them. Thus the collective sum of the forces acting on a particle of mass m will be

$$\vec{F}_g(r) = \sum_i \frac{GmM_i(\vec{r}_i - \vec{r})}{|\vec{r}_i - \vec{r}|^3} = \int_{v'} \frac{Gm\rho(\vec{r}')(\vec{r}' - \vec{r})}{|\vec{r}' - \vec{r}|^3} \, dV' \quad , \tag{4.1.5}$$

where we have included an expression on the right to indicate the total force arising from a continuous mass distribution $\rho(\vec{r})$. Thus the gravitational field resulting from such a configuration is

$$\vec{G}_g(r) = \sum_i \frac{GM_i(\vec{r}_i - \vec{r})}{|\vec{r}_i - \vec{r}|^3} = \int_{v'} \frac{G\rho(\vec{r}')(\vec{r}' - \vec{r})}{|\vec{r}' - \vec{r}|^3} \, dV' \quad . \tag{4.1.6}$$

The potential that will give rise to this force field is

$$\Phi(\vec{r}) = \sum_i \frac{GM_i}{|\vec{r}_i - \vec{r}|} = \int_{V'} \frac{G\rho(\vec{r}')}{|\vec{r}' - \vec{r}|} \, dV' \qquad (4.1.7)$$

The evaluation of the scalar integral of equation (4.1.7) will provide us with the potential (and hence the potential energy of a unit mass) ready for insertion in the Lagrangian. In general, however, such integrals are difficult to do so we will consider a different representation of the potential in the hope of finding another means for its determination.

4.2 Poisson's and Laplace's Equations

The basic approach in this section will be to turn the integral expression for the potential into a differential expression in the hope that the large body of knowledge developed for differential equations will enable us to find an expression for the potential. To do this we will have to make a clear distinction between the coordinate points that describe the location at which the potential is being measured (the field point) and the coordinates that describe the location of the sources of the field (source points). It is the latter coordinates that are summed or integrated over in order to obtain the total contribution to the potential from all its sources. In equations (4.1.5 - 4.1.7) $\vec{r}$ denotes the field points while $\vec{r}'$ labels the sources of the potential.

Consider the Laplacian (i.e., the divergence of the gradient [$(\nabla^2) = (\nabla \cdot \nabla)$]) operating on the integral definition of the potential for continuous sources [the right-most term in equation (4.1.7)]. Since the Laplacian is operating on the potential, we really mean that it is operating on the field coordinates. But the field and source coordinates are independent so that we may move the Laplacian operator through the integral sign in the potential's definition. Thus,

$$\nabla^2 \Phi(\vec{r}) = \int_{V'} G\rho(\vec{r}') \left(\nabla \cdot \nabla \left[\frac{1}{|\vec{r}' - \vec{r}|} \right] \right) dV' \qquad (4.2.1)$$

Since the Laplacian is the divergence of the gradient we may make use of the Divergence theorem

$$\int_V \nabla \cdot \vec{H} \, dV = \int_A \vec{H} \cdot d\vec{A} \qquad (4.2.2)$$

53

to write

$$\int_{V'} G\rho(\vec{r}')[\nabla \cdot \nabla(|\vec{r}'-\vec{r}|^{-1})]dV' = \int_{S} G\rho(\vec{r}')[\nabla(|\vec{r}'-\vec{r}|^{-1})] \cdot d\vec{A}' \quad (4.2.3)$$

Here the surface S is that surface that encloses the volume V'.

Now consider the simpler function (1/r) and its gradient so that

$$\int_{S} \nabla\left[\frac{1}{r}\right] \cdot d\vec{A} = -\int_{S} \frac{\hat{r} \cdot d\vec{A}}{(r)^2} = -\int_{S} d\omega = -\omega \quad . \quad (4.2.4)$$

The integrand of the second integral is just the definition of the differential solid angle so that the integral is just the solid angle ω subtended by the surface S as seen from the origin of r. If the source and field points are different physical points in space, then we may construct a volume that encloses all the source points but does not include the field point. Since the field point is outside of that volume, then the solid angle of the enclosing volume as seen from the field point is zero. However, should one of the source points correspond to the field point, the field point will be completely enclosed by the surrounding volume and the solid angle of the surface as seen from the field point will be 4π steradians. Therefore the integral on the right hand side of equation (4.2.3) will either be finite or zero depending on whether or not the field point is also a source point. Integrands that have this property can be written in terms of a function known as the Dirac delta function which is defined as follows

$$\left.\begin{array}{l} \delta(r) \equiv 0 \text{ for all } r \neq 0 \\ \int \delta(r)dr \equiv 1 \end{array}\right\} \quad . \quad (4.2.5)$$

If we use this notation to describe the Laplacian of (1/r) we would write

$$\nabla^2(1/r) = -4\pi\delta(r) \quad , \quad (4.2.6)$$

and our expression for the potential would become

$$\nabla^2\Phi(r) = \nabla^2\int_{V'} G\rho(\vec{r}')[|\vec{r}'-\vec{r}|]^{-1}dV' = -4\pi G\int_{V'} \delta(r'-r)\rho\vec{r}'dV' . \quad (4.2.7)$$

This integral has exactly two possible results. If the field

54

point is a source point we get

$$\nabla^2 \Phi(r) = -4\pi G \rho(r) \quad , \qquad\qquad (4.2.8)$$

which is known as *Poisson's equation*. If the field point is not a source point, then the integral is zero and we get

$$\nabla^2 \Phi(r) = 0 \quad . \qquad\qquad (4.2.9)$$

This is known as *Laplace's equation* and the solution of either yields the potential required for the Lagrangian and the equations of motion. Entire books have been written on the solution of these equations and a good deal of time is spent in the theory of electrostatics developing such solutions (eg. Jackson[4]). All of that expertise may be borrowed directly for the solution of the potential problem for mechanics.

   In celestial mechanics we are usually interested in the motion of some object such as a planet, asteroid, or spacecraft that does not contribute significantly to the potential field in which it moves. Such a particle is usually called a test particle. Thus, it is Laplace's equation that is of the most interest. Laplace's equation is a second order partial differential equation. The solution of partial differential equations requires "functions of integration" rather than constants of integration expected for total differential equations. These functions are known as boundary conditions and their functional nature greatly complicates the solution of partial differential equations. The usual approach to the problem is to find some coordinate system wherein the functional boundary conditions are themselves constants. Under these conditions the partial differential equations in the coordinate variables can be written as the product of total differential equations which may be solved separately. Such coordinate systems are said to be coordinate systems in which Laplace's equation is separable. It can be shown that there are thirteen orthonormal coordinate frames (see Morse and Feshback[1]) in which this can happen. Unless the boundary conditions of the problem are such that they conform to one of these coordinate systems, so that the functional conditions are indeed constant on the coordinate axes, one must usually resort to numerical methods for the solution of Laplace's equation.

   Laplace's equation is simply the homogeneous form of Poisson's equation. Thus, any solution of Poisson's equation must begin with the solution of Laplace's equation. Having found the homogeneous solution, one proceeds to search for a particular solution. The sum of the two then provides the complete solution for the inhomogeneous Poisson's equation.

   In this book we will be largely concerned with the motion of objects in the solar system where the dominant source of the gravitational potential is the sun (or some planet if one is discussing satellites). It is generally a good first approximation to assume that the potential of the sun and planets is that of a point mass. This greatly facilitates the solution of Laplace's equation and the determination of the potential. However, if one is interested in the motion of satellites about some nonspherical object then the situation is rather more complicated. For the precision required in the calculation of the orbits of spacecraft, one cannot usually assume that the driving potential is that of a point mass and therefore spherically symmetric. Thus, we will spend a little time investigating an alternative method for determining the potential for slightly distorted objects.

4.3 Multipole Expansion of the Potential

Let us return to the integral representation of the gravitational potential

$$\Phi(r) = G \int_V \frac{\rho(r')dV'}{|\vec{r}'-\vec{r}|} \ .$$ (4.3.1)

Assume that the motion of the test particle is such that it never comes "too near" the sources of the potential so that $|\vec{r}'| \ll |\vec{r}|$. Then we may expand the denominator of the integrand of equation (4.3.1) in a Taylor series about r' so that

$$\frac{1}{|r'-r|} \approx \frac{1}{r} + \sum_i r_i' \frac{\partial(1/r)}{\partial r_i} + \frac{1}{2} \sum_i \sum_j r_i' r_j' \frac{\partial^2(1/r)}{\partial r_i \partial r_j}$$

$$-\frac{1}{6} \sum_i \sum_j \sum_k r_i' r_j' r_k' \frac{\partial^3(1/r)}{\partial r_i \partial r_j \partial r_k} \ ,$$ (4.3.2)

or in vector notation

$$[|\vec{r}'-\vec{r}|]^{-1} = (1/r) - [\vec{r}' \cdot \nabla(1/r)] + \frac{1}{2}[\vec{r}'\vec{r}' : \nabla\nabla(1/r)]$$

$$- (1/6)[\vec{r}'\vec{r}'\vec{r}' \vdots \nabla\nabla\nabla(1/r)] \ .$$ (4.3.3)

In chapter 1 we defined the scalar product to represent complete summation over all available indices so that the

56

resulting scalar product of tensors with ranks m and n was |m-n|. However, in order to make clear that multiple summation is needed in equation (4.3.3), I have used multiple "dots". The definition of this notation can be seen from the explicit summation in equation (4.3.2) or can be defined by

$$\vec{AB}:\vec{A'}\vec{B'} \equiv (\vec{A}.\vec{B'})(\vec{B}.\vec{A'}) \quad . \tag{4.3.4}$$

Using this expansion to replace the denominator of the integral definition of the potential [equation (4.3.1)] we get

$$\Phi(r) = G\int_{v'}\frac{\rho(r')dV'}{|\vec{r'}-\vec{r}|} = G\left(\left[\frac{1}{r}\right]\int_{v'}\rho(r')dV' \quad \int_{v'}\vec{r'}\rho(r')dV'.\nabla\left[\frac{1}{r}\right]\right.$$

$$\left. + \frac{1}{2}\int_{v'}\vec{r'}\vec{r'}\rho(r')dV':\nabla\nabla\left[\frac{1}{r}\right] + \cdots + \right) \quad . \tag{4.3.5}$$

This expansion allows the separation of the dependence of the field coordinates from the source coordinate. Thus the integrals are properties of the source of the potential only and may be calculated separately from any other aspect of the mechanics problem. Once known, they give the potential explicitly as a function of the field coordinates alone and this is what we need for specifying the Lagrangian. We can make this clearer by re-writing equation (4.3.5) as

$$\Phi(r) = \{M(1/r) - \vec{P}.\nabla(1/r) + \frac{1}{2}Q:\nabla\nabla(1/r)$$

$$- (1/6)\underline{S}\vdots\nabla\nabla\nabla(1/r) + \cdots + \}. \tag{4.3.6}$$

This expansion of the potential is known as a "multipole" expansion for the parameters $M,\vec{P},Q$, and $\underline{S}$ which are known as the multipole moments of the source distribution. For the gravitational potential the unipole moment is a scalar and just equal to the total mass of the sources of the potential. The vector quantity $\vec{P}$ is called the dipole moment and Q is the tensor quadrupole moment, etc. The higher order moments are in turn higher order tensors. The repeated operation of the del-operator ∇ on the quantity $(1/r)$ also produces higher order tensors which are simply geometry and have nothing to do with the mass distribution itself. The first two of these are

$$\left.\begin{array}{l}\nabla(1/r) = -\ \hat{r}/r^2 \\[2mm] \nabla\nabla(1/r) = -\ [\mathbf{1}-3\hat{r}\hat{r}]/r^3\end{array}\right\} \quad . \tag{4.3.7}$$

As one considers higher order terms the geometrical tensors

57

represented by the multiple gradient operators contain a larger and larger inverse dependence on r and therefore play a successively diminished role in determining the potential. Thus we have effectively separated the positional dependence of the field point from the mass distribution that produces the various multipole moments.

By way of example, let us consider two *unequal* mass points separated by a distance ℓ, located on the z-axis, and with the coordinate origin at the center of mass (see Figure 4.1). From the definition of the multipole moments, we have

$$\left.\begin{aligned}
M &\equiv \int_V \rho(r')dV' = m_1 + m_2 \\
\vec{P} &\equiv \int_V \vec{r}'\rho(r')dV' = [m_1 m_2/(m_1+m_2) - m_2 m_1/(m_1+m_2)]\ell\,\hat{k} = 0 \\
\underline{Q} &\equiv \int_V \vec{r}'\vec{r}'\rho(r')dV' = [m_1 m_2/(m_1+m_2)]\ell^2\,\hat{k}\hat{k} \\
\underline{S} &\equiv \int_V \vec{r}'\vec{r}'\vec{r}'\rho(r')dV' = (m_1 z_1^3 + m_2 z_2^3)\,\hat{k}\hat{k}\hat{k}
\end{aligned}\right\} (4.3.8)$$

which when combined with the coordinate representation of equation (4.3.6) yields a series expansion for the potential of the form

$$\Phi(r) = G\{(m_1+m_2)/r$$

$$+ (\ell^2/2)[m_1 m_2/(m_1+m_2)](1-3\cos^2\theta)/r^3 + \cdots + . \quad (4.3.9)$$

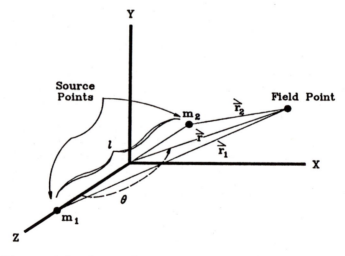

Figure 4.1 shows the arrangement of two unequal masses for the calculation of the multipole potential resulting from them.

58

Unless the field point comes particularly close to the sources, this series will converge quickly. We can also make use of a pleasant property of the gravitational force, namely that there are no negative "charges" in the force law of gravitation. Thus we may always choose a coordinate system such that

$$\vec{P}(\vec{r}') = \int_V \vec{r}' \rho(r') dV' = 0 \quad . \tag{4.3.10}$$

This means that for celestial mechanics there will never be a dipole moment of the potential as long as we choose the coordinate frame properly. This is usually done by taking advantage of any symmetry presented by the object and locating the origin at the center of mass. Not only does the dipole moment vanish, but for objects exhibiting plane symmetry all even moments of the multipole expansion vanish for the gravitational potential. This certainly enhances the convergence of the series expansion for the potential and means that the first term that must be included after the point-mass potential term is the quadrupole term. The inclusion of this term means that the error in the potential will be of the order $O(1/r^5)$. Even though the potential represented by a multipole expansion converges rapidly with increasing distance, the contribution of such terms can be significant for small values of r. Thus there is great interest in determining the magnitude of these terms for the potential field of the earth so that the orbits of satellites may be predicted with greater certainty.

We have now described methods whereby the potential can be calculated for an arbitrary collection of mass points to an arbitrary degree of accuracy. The insertion of the potential into the Lagrangian will enable one to determine the equations of motion and the solution of these equations then constitutes the solution of any classical mechanics problem. Therefore, let us now turn to the solution of specific problems found in celestial mechanics.

Chapter 4: Exercises

1. The potential energy of the interaction between a multipole T and a scalar potential field Φ is given by

$$U = T^{(i)} \odot \nabla^{(i)} \Phi \quad ,$$

where $T^{(i)}$ is a tensor of rank (i) and $\nabla^{(i)}\Phi$ describes i applications of the del operator to the scalar potential Φ. The symbol $\odot$ stands for the most general application of the scalar product, namely the contraction (i.e., the addition) of the two resulting tensors over all indices.

   a: Consider four equal masses with cartesian three dimensional coordinates

mass #	X	Y	Z
#1	-1	0	0
#2	+1	0	0
#3	0	2	2
#4	0	-3	-3

   Find the total self energy of the system.
   b: Find the potential energy of the above system with a fifth identical mass located at (0,0,0).

2. Given that the interaction energy of a dipole and quadrupole may be written as

$$U_{pq} = \bar{P} \cdot \nabla \Phi_Q \quad \Bigg)$$

   or

$$U_{qp} = Q : \nabla \nabla \Phi_P \quad \Bigg) \quad ,$$

   show that $U_{pq} = U_{qp}$.

3. Use a multipole expansion to find the potential field of three equal mass points located at the vertices of an equilateral triangle with side d. Restrict your solution to the plane of the triangle and keep only the first two terms of the expansion.

4. Find the interaction energy of a 10 kg sphere with the Earth-Moon system when the three are located so as to form an equilateral triangle. Assume the Earth and Moon are spherical. Compare the relative importance of the first two terms of the multipole expansion for the Earth-Moon potential.

5

Motion under the Influence of a Central Force

        The fundamental forces of nature depend only on the distance from the source. All the complex interactions that occur in the real world arise from these forces, and while many of them are usually described in a more complex manner, their origin can be found in the fundamental forces that depend only on distance. Thus even the intricate forces of aerodynamic drag can ultimately be described as resulting from the electrostatic potential of the air molecules scattering with the electrostatic potential of the molecules of the aircraft, and the electrostatic force is essentially one that depends on distance alone. It is the presence of many sources of the distance-dependent forces that enables the complex world we know to exist. Thus, in order to understand complex phenomena, it is appropriate that we begin with the simplest. Therefore we will begin by applying the tools of mechanics developed in the previous two chapters to describe the motion of an object moving under the influence of a single source of a force that depends only on the distance. We will call this object a "test particle" to make clear that its motion in no way affects the source of the potential. Such a situation is known as a central force problem since the source may be located at the origin of the coordinate system making it central to the resulting description.

5.1 Symmetry, Conservation Laws, the Lagrangian, and Hamiltonian for Central Forces

        Since there is a single source producing a force that

depends only on distance, the force law is spherically symmetric. If this is the case, then there can be no torques present in the system as there would have to be a preferred axis about which the torques act. That would violate the spherical symmetry so

$$\vec{N} = d\vec{L}/dt = 0. \qquad (5.1.1)$$

Equation (5.1.1) clearly means that the total angular momentum of the test particle does not change in time. Specifically, it means that the direction of the angular momentum vector doesn't change. Since there is only one particle in this system, this is little more than a statement of the conservation of angular momentum, but it has a great simplifying implication. The radius vector $\vec{r}$ and the particle's linear momentum $\vec{p}$ define a plane. Since

$$\vec{L} = \vec{r} \times \vec{p} \quad , \qquad (5.1.2)$$

the angular momentum is always perpendicular to that plane and being constant in space requires that the motion of the particle is confined to that plane. Thus we can immediately reduce the problem to a two dimensional description.

Since there is only one particle in the system and we require the total energy of the system to be constant, the total energy of the particle must be constant. Thus such a force is conservative and we may use the Lagrangian formalism of chapter 3 to obtain the equations of motion. We begin this procedure by choosing a set of generalized coordinates. Remember that the only requirement for the generalized coordinates is that they span the space of the motion and be linearly independent. For motion that is confined to a plane defined by the action of a central force, the logical choice of a coordinate frame is polar coordinates with the center of the force field located at the origin of the coordinate system. However, since the kinetic energy is more obviously written in cartesian coordinates, let us use the definition of the Lagrangian to write

$$\mathcal{L} \equiv T - V = \tfrac{1}{2}m[\dot{x}^2 + \dot{y}^2] - m\Phi(r) \quad , \qquad (5.1.3)$$

where $\Phi(r)$ is the potential giving rise to the conservative central force. The transformation from cartesian coordinates to polar coordinates is

$$\left. \begin{array}{l} x = r\cos\theta \\[2mm] y = r\sin\theta \end{array} \right\} \quad , \qquad (5.1.4)$$

so that

$$\dot{x}^2 = \dot{r}^2\cos^2\theta - 2r\dot{r}\,\sin\theta\,\cos\theta\,\dot{\theta} + r^2\dot{\theta}^2\,\sin^2\theta \qquad \Bigg\}$$
$$\dot{y}^2 = \dot{r}^2\sin^2\theta + 2r\dot{r}\,\sin\theta\,\cos\theta\,\dot{\theta} + r^2\dot{\theta}^2\,\cos^2\theta \qquad . \tag{5.1.5}$$

Substitution of these expressions into the Lagrangian gives

$$\mathcal{L} = \tfrac{1}{2}m[\dot{r}^2 + r^2\dot{\theta}^2] - m\Phi(r) \qquad . \tag{5.1.6}$$

Lagrange's equations of motion for polar coordinates will then be

$$\frac{d}{dt}\left[\frac{\partial\mathcal{L}}{\partial\dot{r}}\right] - \frac{\partial\mathcal{L}}{\partial r} = 0 \qquad \Bigg\}$$
$$\frac{d}{dt}\left[\frac{\partial\mathcal{L}}{\partial\dot{\theta}}\right] - \frac{\partial\mathcal{L}}{\partial\theta} = 0 \qquad . \tag{5.1.7}$$

In terms of the polar coordinates $[r,\theta]$ the quantities required for Lagrange's equations of motion are

$$\frac{\partial\mathcal{L}}{\partial\dot{r}} = m\dot{r}$$

$$\frac{\partial\mathcal{L}}{\partial\dot{\theta}} = mr^2\dot{\theta}$$

$$\frac{\partial\mathcal{L}}{\partial r} = mr\dot{\theta}^2 - m\frac{\partial\Phi(r)}{\partial r} \qquad \Bigg\}$$

$$\frac{\partial\mathcal{L}}{\partial\theta} = 0 \tag{5.1.8}$$

so that the explicit equations of motion become

$$m\ddot{r} - mr\dot{\theta}^2 + m\frac{\partial\Phi(r)}{\partial r} = 0 \qquad \Bigg\}$$

$$\frac{d}{dt}(mr^2\dot{\theta}) = 2mr\dot{r}\dot{\theta} + mr^2\ddot{\theta} = 0 \tag{5.1.9}$$

63

Now in chapter 3 we developed the Hamiltonian from the Lagrangian and the generalized momenta [see equations (3.3.1-3.3.3)] and it is illustrative to do this explicitly for the case of a central force. The generalized momenta can be obtained from the Lagrangian by means of equation (3.3.1) so for polar coordinates we have

$$\left. \begin{array}{l} q_i = [r, \theta] \\ \\ p_i = [m\dot{r}, mr^2\dot{\theta}] \end{array} \right\} \qquad . \qquad (5.1.10)$$

From equation (3.3.3) we can then write the Hamiltonian as

$$H(p_r, p_\theta, r, \theta, t) = [m\dot{r}^2 + mr^2\dot{\theta}^2] - [\tfrac{1}{2}m\dot{r}^2 + \tfrac{1}{2}mr^2\dot{\theta}^2] + m\Phi(r)$$

$$= T + U = E \quad . \qquad (5.1.11)$$

As long as the Lagrangian and generalized coordinates were not explicit functions of time, the Hamiltonian is an integral of the motion. Since from equation (5.1.11) it is clear that the Hamiltonian is also the total energy, we have effectively recovered the law of conservation of energy. The Hamilton canonical equations of motion [see equations (3.3.6)] effectively add nothing new to the problem since we already have two constants of the motion (i.e., the angular momentum and the total energy). Thus we can turn to the implications of these two constants

5.2 The Areal Velocity and Kepler's Second Law

The θ-equation of motion that gives us the constancy of angular momentum enables us to write

$$mr^2\dot{\theta} = mr\omega = L \quad . \qquad (5.2.1)$$

The differential area included between two radius vectors separated by a differential angle $d\theta$ is just

$$dA = \tfrac{1}{2}r(rd\theta) \quad . \qquad (5.2.2)$$

Let us call the time derivative of this area the areal velocity so that

$$\frac{dA}{dt} = \tfrac{1}{2}r^2\frac{d\theta}{dt} = \tfrac{1}{2}L/m = \text{const.} \qquad (5.2.3)$$

Thus we could say that the areal velocity of the test particle is constant and merely be re-stating the conservation of angular momentum. This is indeed the way Johannes Kepler gave his second law of planetary motion. However, Kepler had no conception of angular momentum and his laws dealt only with the planets. Here we see that Kepler's second law not only applies to objects moving under the influence of the gravitational force, but will hold for an object moving under the influence of any central force regardless of its distance dependence. Thus Kepler's second law of planetary motion is far more general than Kepler ever knew.

We may use this result to eliminate $\dot{\theta}$ from the first of the two Lagrangian equations of motion and thereby reduce the problem to that of one dimension. Solving equation (5.2.1) for $\dot{\theta}$ and substituting into the r-equation (5.1.9) we get

$$m\ddot{r} - L^2/(mr^3) + \partial\Phi(r)/\partial r = 0 \quad . \tag{5.2.4}$$

We can obtain the result of the Hamiltonian directly by multiplying equation (5.2.4) by $\dot{r}$ and re-writing as

$$\frac{d}{dt}[\tfrac{1}{2}m\dot{r}^2] = -\frac{d}{dt}[\Phi(r)+L^2/2mr^2]m = -\frac{d}{dt}[\Phi(r)+\tfrac{1}{2}r^2\dot{\theta}^2]m \quad , \tag{5.2.5}$$

which becomes

$$\frac{d}{dt}[\tfrac{1}{2}m(\dot{r}^2+r^2\dot{\theta}^2)+m\Phi(r)] = \frac{d}{dt}[T+V] = \frac{dE}{dt} = 0 \tag{5.2.6}$$

5.3 The Solution of the Equations of Motion

Finding the two integrals of the motion goes a long way to completing the solution of the problem. These two constants essentially amount to integrating each of equations (5.1.9) once so that the equations of motion can be written as

$$\left.\begin{array}{l}\tfrac{1}{2}m\dot{r}^2 + [L^2/(2mr^2) + m\Phi(r)] = E = \text{const.} \\[2mm] mr\dot{\theta} = L = \text{const.}\end{array}\right\} \tag{5.3.1}$$

The first of these is obtained by integrating equation (5.2.6) and replacing $\dot{\theta}$ from equation (5.2.1). The other is equation (5.2.1) itself. These are two first order differential equations and require two more constants to completely specify their solution. These two additional constants are known as initial conditions and it is worth distinguishing them from integrals of the motion. An *integral of the motion* will have the same value for the entire temporal history of the system while an *initial value* or *boundary condition* is just the value of one of the dependent variables of the problem at some *specific* instant in time. Surely the latter is a constant since it is specified at a given time, but an integral of the motion is some combination of the dependent variables that is constant for all time. Integrals of the motions are exceedingly important to any dynamics problem and knowledge of them, as we shall see later, places very useful constraints on the history and nature of the system.

The solution of the second of equations (5.3.1) yields the temporal history of the variable θ and can be obtained by direct integration of that equation so that

$$\theta(t) = \int_0^t \frac{L dt}{mr^2(t)} + \theta_0 \quad . \tag{5.3.2}$$

Here the constant θ_0 is the initial value of θ at $t = 0$. We have chosen the initial value of the time to be zero, but that is arbitrary and the initial value could have been anything. The first of equations (5.3.1) is somewhat more difficult to solve. Direct integration gives

$$t = \left[\frac{m}{2}\right]^{\frac{1}{2}} \int_{r_0}^r \frac{dr}{[E - \Phi(r) - L^2/2mr^2]^{\frac{1}{2}}} = t(r) \quad . \tag{5.3.3}$$

Thus the r-coordinate is given as an implicit function of time $t(r)$. This implicit function must be inverted to have the same form as equation (5.3.2). The parameter r_0 is the second initial value, being the value of r at $t = 0$. Thus the two initial values $[\theta_0, r_0]$, and the integrals of the motion $[L, E]$ completely specify the motion of the particle. However, to solve a specific problem we must specify $\Phi(r)$ because the integral, and subsequent inversion of equation (5.3.3), cannot be done without knowledge of $\Phi(r)$.

In any event, we may put rather general limits on the range of solutions that we can expect for any given $\Phi(r)$. The second of equations (5.3.3) is essentially a one dimensional equation in r, so we will define a new potential

$$\emptyset(r) = \Phi(r) + L^2/2m^2r^2 \quad . \tag{5.3.4}$$

We can require that the potential energy vanish as $r \to \infty$ as a reasonable boundary condition. Thus the kinetic energy is

$$\tfrac{1}{2}m\dot{r}^2 = E - m\emptyset(r) \quad . \tag{5.3.5}$$

66

Now if $\emptyset(r) \geq 0$ then the force law is repulsive, and there is some minimum distance r_{min} to which the particle can approach the source before the right hand side of equation (5.3.5) would become negative, implying a nonphysical negative kinetic energy. This is a plausible result that simply says that a repulsive force which increases in strength as its source is approached will eventually stop the approach. The total energy for such a system must also always be greater than zero.

For the more interesting case of power law potentials where $\emptyset(r) \leq 0$, we can write

$$\emptyset(r) = [\tfrac{1}{2}L^2 - k/r^{(n-2)}]/r^2 \quad , \tag{5.3.6}$$

where k is some positive constant. Interesting things will happen at

$$r_0 = (2mk/L^2)^{[1/(n-2)]} \quad . \tag{5.3.7}$$

If power law dependence is such that $n > 2$, then r_0 serves as an upper bound for particles with a total energy $E < 0$.

For the more pertinent case of gravity where $n = 1$, then r_0 serves as a lower bound inside of which particles may not approach. The physical interpretation of this lower bound is simple. If L is not zero, then there is some angular motion of the particle as it orbits the central source. However, as it approaches the central source, the conservation of angular momentum will require an increase in its angular velocity keeping the particle from approaching closer. Colloquially one could say that the particle is repelled by centrifugal force. Indeed, this part of the pseudo-potential $\emptyset$ is often known as the "rotational potential". Thus, if the total angular momentum $L > 0$, then the motion of the particle will be kept beyond r_0. If the total energy $E < 0$, then there will also be an upper bound r_1 since the kinetic energy must remain positive. Thus for cases in which the total energy $E < 0$, the particle's motion will be confined between $r_1 \geq r \geq r_0$. This is the case for virtually all motion in the solar system.

5.4 The Orbit Equation and Its Solution for the Gravitational Force

To see more clearly the solution to the equations of motion, let us eliminate time as the independent variable from equations (5.1.9). Thus, we search for a single equation involving r as a function of θ. We can do this by noting that the total time derivative operator can be written as

67

$$\frac{d}{dt} = \frac{d\theta}{dt}\frac{d}{d\theta} = \frac{L}{mr^2}\frac{d}{d\theta} \quad , \tag{5.4.1}$$

and

$$\frac{1}{r^2}\frac{dr}{d\theta} = -\frac{d(1/r)}{d\theta} \quad . \tag{5.4.2}$$

Thus, replacing the time derivatives of equation (5.1.9) by equation (5.4.1) and using equation (5.4.2) we can write

$$\frac{-L^2}{mr^2}\frac{d^2(1/r)}{d\theta^2} - \frac{L^2}{mr^3} = -m\frac{\partial\Phi(r)}{\partial r} \equiv f(r) \quad . \tag{5.4.3}$$

This is the so called orbit equation since its solution is $r(\theta)$, the orbit of the particle. This equation is more amenable to solution if we re-write it by substituting

$$u \equiv 1/r \quad , \tag{5.4.4}$$

so that

$$\frac{L^2 u^2}{m}\left[\frac{d^2 u}{d\theta^2} + u\right] = -f(1/u) \quad . \tag{5.4.5}$$

The quantity $f(r)$ or $f(1/u)$ is the *force* law, which for gravity is

$$f(r) = -GMm/r^2 = -GMmu^2 \quad . \tag{5.4.6}$$

Thus the orbit equation for the gravitational force takes the following relatively simple form:

$$\frac{d^2 u}{d\theta^2} + u = \frac{GMm^2}{L^2} = \beta = \text{const.} \tag{5.4.7}$$

This is a second order equation so we can expect two constants of integration in the most general solution. As is customary with differential equations of this form, we can guess a solution to be

$$u = A\cos(\theta - \theta_0) + \beta \quad , \tag{5.4.8}$$

which we can re-write as

68

$$r = P/[1+e \, \cos(\theta-\theta_0)] \quad , \qquad\qquad (5.4.9)$$

so long as

$$\left. \begin{array}{l} P = (1/\beta) = L^2/GMm^2 \\[2ex] e = AL^2/GMm^2 \end{array} \right\} \qquad\qquad (5.4.10)$$

The reason for this last transformation is that equation (5.4.9) is the general equation for a conic section with a focus at the coordinate origin and an eccentricity e. Thus we recover the essence of Kepler's first law, namely that the planets move in ellipses with the sun at one focus. Certainly ellipses are conic sections, and should the mass of the sun greatly exceed the mass of the earth, then the sun may be regarded as the source of the central force of gravity. We have now only to decide which type of conic section the orbit will be and on what parameter the kind will depend.

We can make this determination by generating the total energy from the solution to the orbit equation. First differentiate the solution with respect to time and after some algebra find the velocity to be given by

$$v^2 = [L^2/m^2 P^2][(e^2-1)+2P/r] \quad . \qquad\qquad (5.4.11)$$

Now form the kinetic energy and add the potential energy to get

$$E = \tfrac{1}{2}(GMm/P)(e^2-1) \quad . \qquad\qquad (5.4.12)$$

It is clear from equation (5.4.12) that the sign of the total energy will determine the sign of the eccentricity e and hence the type of orbit namely

$$\left. \begin{array}{l} E > 0 \Rightarrow e^2 > 1 : \text{hyperbolic orbit} \\[2ex] E = 0 \Rightarrow e^2 = 1 : \text{parabolic orbit} \\[2ex] E < 0 \Rightarrow e^2 < 1 : \text{elliptic orbit} \end{array} \right\} \qquad (5.4.13)$$

Thus we see that Kepler's first law of planetary motion implies that the total orbital energy of the planets is negative so that the planets are bound. In addition, any bound test particle in orbit about the sun will have an elliptic orbit.

Chapter 5: Exercises

1. Given that a particle is moving under the influence of a central force of the form

$$f = -\frac{k}{r^2} + \frac{c}{r^3} \quad ,$$

where k and c are positive constants. Show that the solution to the orbit equation can be put in the form

$$r = \frac{a(1-e^2)}{1-e\,\cos(\alpha\theta)} \quad ,$$

which is an ellipse for $e < 1$ and $\alpha = 1$. Discuss the character of the orbit for $\alpha \neq 1$ and $e < 1$. Derive an approximate expression for α in terms of the dimensionless parameter $\gamma = [c/(ka)]$.

2. Discuss the motion of a test particle moving in a potential field of the form

$$\Phi(r) = (\alpha/r) + (\beta/r^3) \quad ,$$

in terms of the rotational potential and conservation of energy.

3. A particle moves in a circular orbit of radius r_0 under the influence of a central force located at some point inside the orbit. The minimum and maximum speeds of the particle are v_1 and v_2 respectively. Find the orbital period in terms of these speeds and the radius of the orbit.

4. Suppose that all the planets move about the sun in circular orbits under the influence of an inverse cube force law. Assuming conservation of momentum and energy, find a relation between the orbital period and the radius of the orbit for the planets. (i.e., a new "Kepler's third law").

6

The Two Body Problem

The classical problem of celestial mechanics, perhaps of all Newtonian mechanics, involves the motion of one body about another under the influence of their mutual gravitation. In its simplest form, this problem is little more than the generalization of the central force problem, but in some cases the bodies are of finite size and are not spherical. This may complicate the problem immensely as the potential fields of the objects no longer vary as the inverse square of the distance. This causes orbits to precess and the objects themselves to undergo gyrational motion. This latter motion results from external torques produced on a nonspherical object interacting with the object's own spin angular momentum. While we will not deal with the more difficult aspects of these phenomena in this book, it is useful to understand something of the properties of finite rigid bodies so that we are equipped to begin to understand some of the difficulties when they arise. Thus, we will begin our discussion of the two body problem with a summary of the properties of rigid bodies.

6.1 The Basic Properties of Rigid Bodies

Let us begin by assuming that the rigid object we are considering is located in some orthonormal coordinate system so that the points within the object can be located in terms of some vector $\vec{r}$.

a. The Center of Mass and the Center of Gravity

Let us define two concepts usually taken for granted in mechanics books. First the *center of mass* is simply a 'mass weighted' mean position for the object. Again I will give both the discrete and continuous forms so that

$$r_c = \sum_i m_i \vec{r}_i / \sum_i m_i = \int_V \vec{r} \rho(\vec{r}) dV / M \quad . \tag{6.1.1}$$

A second concept that is often confused with the center of mass is the *center of gravity*. This is often defined to be that point where the force of gravity can be considered to be acting. Mathematically that would mean that all torques produced by gravity would vanish about that point so that

$$\vec{r}_g \times \sum_i \vec{f}_i = \sum_i \vec{r}_i \times \vec{f}_i = \int_V [\vec{r} \times \rho(\vec{r}) \vec{g}] dV = 0 \quad . \tag{6.1.2}$$

In a cartesian coordinate frame this could be expressed in coordinate form as

$$\left.\begin{aligned}
r_{2,g} A_3 - r_{3,g} A_2 &= B_{2,3} - B_{3,2} \\
r_{3,g} A_1 - r_{1,g} A_3 &= B_{3,1} - B_{1,3} \\
r_{1,g} A_2 - r_{2,g} A_1 &= B_{1,2} - B_{2,1}
\end{aligned}\right\} \tag{6.1.3}$$

where

$$\left.\begin{aligned}
A_j &= \sum_i g_{ij} m_i \\
B_{ij} &= \sum_i r_{ik} g_{ij} m_i
\end{aligned}\right\} \quad . \tag{6.1.4}$$

If one writes this as a linear system of equations for the components of the vector defining the center of gravity one gets

$$A r_g = \begin{pmatrix} 0 & A_3 & -A_2 \\ A_3 & 0 & A_1 \\ -A_2 & A_1 & 0 \end{pmatrix} \begin{pmatrix} r_{1,g} \\ r_{2,g} \\ r_{3,g} \end{pmatrix} = \begin{pmatrix} B_{2,3} - B_{3,2} \\ B_{3,1} - B_{1,3} \\ B_{1,2} - B_{2,1} \end{pmatrix} \quad . \tag{6.1.5}$$

However,

$$\text{Det } \mathbf{A} = 0. \tag{6.1.6}$$

This means that the equations are singular and there is no
unique definition, so that the magnitude of r_g is undefined.
Only if we require that $|\vec{r}_g|=|\vec{r}_c|$ and that the gravity vector
be constant can we define a unique vector which will be equal
to the vector to the center of mass. Thus, if the gravity field
varies over the object, the center of gravity is not uniquely
defined. In the case in which it is well defined it is the same
as the center of mass. Physically one can see this by imagining
all the points within a body where one could attach a hook,
suspend the object and not have it move. Any such points would
serve as the center of gravity. The problem arises from the
cross product and the definition. If one adds to the standard
definition that the center of gravity is that point about which
all the gravitational torques vanish *regardless of the
orientation of the body with respect to the gravitational
field*, then the definition is more tractable.

b. The Angular Momentum and Kinetic Energy about
   the Center of Mass

Consider that the object is rotating about some point
that is fixed with respect to an inertial coordinate frame
(i.e., one that has no accelerative motions). Then the angular
momentum of the object will just be

$$\vec{L} = \sum_i m_i (\vec{r}_i \times \vec{v}_i) = \int_V \rho(\vec{r})(\vec{r} \times \vec{v}) dV \quad , \tag{6.1.7}$$

where

$$\vec{v}_i = \vec{\omega} \times \vec{r}_i \quad . \tag{6.1.8}$$

Since we are considering the object to be rigid, then all
points within the body will rotate with the same *angular
velocity* ω. If that were not true some points within the body
would catch up with others while moving away from still others
and we would not call the body rigid. This allows us to
separate the rotational motion from the positions of points
within the object. Thus by making use of the vector identities
from chapter 1 we may write the angular momentum of the object
as

$$\vec{L} = \sum_i m_i [\vec{r}_i \times (\vec{\omega} \times \vec{r}_i)] = \sum_i m_i [\vec{\omega} r_i^2 - \vec{r}_i (\vec{r}_i \cdot \vec{\omega})] \quad . \tag{6.1.9}$$

Writing out equation (6.1.9) for each component of $\vec{L}$ we see
that equation (6.1.9) can be re-written as

$$\vec{L} = I.\vec{\omega} \quad , \tag{6.1.10}$$

where I is known as the moment of inertia tensor and has components

$$I_{jk} = \left\{ \begin{array}{ll} \sum\limits_{i} m_i (r_i^2 - x_k^2) & \text{for } j = k \\ \sum\limits_{i} m_i x_j x_k & \text{for } j \neq k \end{array} \right\} \tag{6.1.11}$$

Now the kinetic energy of a rotating object about some fixed point is just

$$T = \tfrac{1}{2}\sum\limits_{i} m_i v_i^2 = \tfrac{1}{2}\int_V \rho(\vec{r}) v^2(\vec{r}) dV = \tfrac{1}{2}\int_V \rho(\vec{r}).(\vec{\omega}\times\vec{r}) dV \quad . \tag{6.1.12}$$

Making use of the so-called vector triple product

$$\vec{A}.(\vec{B}\times\vec{C}) = (\vec{A}\times\vec{B}).\vec{C} = \vec{C}.(\vec{A}\times\vec{B}) \quad , \tag{6.1.13}$$

we can write this as

$$T = \tfrac{1}{2}\vec{\omega}.\int_V \rho(\vec{r})(\vec{r}\times\vec{v}) dV = \tfrac{1}{2}\vec{\omega}.\vec{L} \quad . \tag{6.1.14}$$

This can be expressed in terms of the moment of inertia tensor by replacing the angular momentum with equation (6.1.10) so that

$$T = \tfrac{1}{2}\vec{\omega}.I.\vec{\omega} = \tfrac{1}{2}\omega^2 [\hat{n}.I.\hat{n}] = \tfrac{1}{2}\omega^2 I \quad . \tag{6.1.15}$$

Here $\hat{n}$ is a unit vector pointing in the direction of the angular velocity vector and the quantity in square brackets is then just a property of the body and is called the moment of inertia about the axis $\hat{n}$. Clearly the moment of inertia tensor, I, will have the symmetric property

$$I_{ij} = I_{ji} \quad . \tag{6.1.16}$$

c. The Principal Axis Transformation

Calculations involving the moment of inertia tensor would be a lot easier if there were some coordinate frame in which the tensor were diagonal. It is clear from equation (6.1.11) that the tensor is a symmetric tensor so that the off diagonal terms satisfy

$$I = [\hat{n}.I.\hat{n}] = \int_V \rho(\vec{r}) [r^2 - \vec{r}.\hat{n}] dV \quad . \tag{6.1.17}$$

74

Thus in order to make the tensor diagonal we need only transform to a coordinate frame wherein the off-diagonal elements are zero. We saw in chapter 2 that one could reach any orthonormal coordinate frame from any other through a series of three coordinate rotations about the successive coordinate axes. This is represented by three independent parameters in the transformation (i.e., the rotation angles). Since we have three constraints to meet (i.e., making the off-diagonal elements zero), it is clear that this can be done. Another way of visualizing this transformation is to scale the unit vector $\hat{n}$ by $\sqrt{I}$ so that

$$\vec{\xi} = \sqrt{I}\hat{n} \quad . \tag{6.1.18}$$

In terms of the components of this vector the expression for the moment of inertia given by equation (6.1.17) becomes

$$I_{11}\xi_1^2 + I_{22}\xi_2^2 + I_{33}\xi_3^2 + I_{12}\xi_1\xi_2 + I_{13}\xi_1\xi_3 + I_{23}\xi_2\xi_3 = 1, \tag{6.1.19}$$

which is the general equation for an ellipsoid. Now there always is a coordinate frame aligned with the principal axes of the ellipsoid where the general equation for the surface becomes

$$I_1'(\xi_1')^2 + I_2'(\xi_2')^2 + I_3'(\xi_3')^2 = 1 \quad . \tag{6.1.20}$$

This coordinate system is known as the *principal axis* coordinate system and it is the coordinate frame in which the off-diagonal elements of the moment of inertia tensor vanish. The diagonal elements are known as the principal moments of inertia, as they are indeed the moments of inertia about the principal axes. They are basically the eigenvalues of the moment of inertia tensor and so can be found from the determinental equation

$$\mathrm{Det}\begin{vmatrix} (I_{11}-I) & I_{12} & I_{13} \\ I_{21} & (I_{22}-I) & I_{23} \\ I_{31} & I_{32} & (I_{33}-I) \end{vmatrix} = 0 \quad , \tag{6.1.21}$$

which is nothing more that a polynomial in I. The principal moments of inertia are the roots of that polynomial.

The moment of inertia is an important concept if one is interested in the motion of an object. For example, it is essential for the understanding of precession. In the rotational equations of motion for an object the moment of inertia plays the role taken by the mass in the dynamical equations of motion of a system of particles.

75

6.2 The Solution of the Classical Two Body Problem

In principle we have assembled all the tools and concepts needed to solve some very difficult mechanics problems. To illustrate the methods needed to determine planetary motion we will consider the classical two body problem of celestial mechanics. We know immediately that we will have two second order vector differential equations to solve for the motion of both objects. Each of these equations will require six independent constants to specify the complete solution. Therefore we may expect to have to find a total of twelve constants of the motion before we can consider the problem solved.

a. The Equations of Motion

In order to find the equations of motion for two bodies moving under their mutual gravity we shall follow much the same procedure that we did for a central force. In order to keep the problem simple we will further assume that the potential of each body is that of a point mass m_1 and m_2 respectively. The kinetic and potential energies of the system are then

$$\left. \begin{array}{l} T = \tfrac{1}{2}m_1(\dot{\vec{r}}_1 \cdot \dot{\vec{r}}_1) + \tfrac{1}{2}m_2(\dot{\vec{r}}_2 \cdot \dot{\vec{r}}_2) \\[2ex] V = Gm_1m_2/|\vec{r}_1 - \vec{r}_2| \end{array} \right\} \quad , \qquad (6.2.1)$$

where $\vec{r}_1$ and $\vec{r}_2$ are position vectors to the objects. These vectors are linearly independent so they form a suitable set of generalized coordinates in which to formulate the Lagrangian equations of motion. Now the elements that enter into the Lagrangian equations of motion are

$$\left. \begin{array}{l} \partial \mathcal{L}/\partial \dot{\vec{r}}_i = m_i \dot{\vec{r}}_i \\[2ex] \partial \mathcal{L}/\partial \vec{r}_i = \partial V/\partial r_i = -Gm_1m_2(\vec{r}_i - \vec{r}_j)/d_{ij}^3 \end{array} \right\} \quad , \qquad (6.2.2)$$

where

$$d_{ij} \equiv |\vec{r}_i - \vec{r}_j| \quad . \qquad (6.2.3)$$

This leads to two vector equations of motion for the two bodies:

$$\left. \begin{array}{l} m_1 \ddot{\vec{r}}_1 + Gm_1m_2(\vec{r}_1 - \vec{r}_2)/d^3 = 0 \\[2ex] m_2 \ddot{\vec{r}}_2 + Gm_1m_2(\vec{r}_2 - \vec{r}_1)/d^3 = 0 \end{array} \right\} \quad . \qquad (6.2.4)$$

76

If we add these equations we get

$$m_1 \ddot{\vec{r}}_1 + m_2 \ddot{\vec{r}}_2 = 0 \quad , \tag{6.2.5}$$

which can be integrated immediately twice with respect to time
to yield

$$m_1 \vec{r}_1 + m_2 \vec{r}_2 = \vec{A}t + \vec{B} \quad . \tag{6.2.6}$$

Note that $\vec{A}$ and $\vec{B}$ are vectors and so contain six linearly
independent constants. From the definition of the center of
mass [equation (6.1.1)] we can write

$$M\vec{r}_c = \vec{A}t + \vec{B} \quad , \tag{6.2.7}$$

which says that at time $t = 0$ the center of mass was located at
$\vec{B}/M$ and was moving with a uniform velocity $\vec{A}/M$. Thus we have
immediately found six of the twelve constants of the motion.
They are the location and velocity of the center of mass.

Since a coordinate frame that undergoes uniform motion is
an inertial coordinate frame (i.e., no accelerations) the laws
of physics will look the same in a coordinate frame moving with
the center of mass as they did in our initial coordinate
system. Therefore we will transform to an inertial coordinate
frame with the origin located at the center of mass. In such a
coordinate system

$$m_1 \vec{r}_1' + m_2 \vec{r}_2' = 0 \quad . \tag{6.2.8}$$

We may use this constraint to decouple each of equations
(6.2.4) from the other so that

$$\left. \begin{array}{l} \ddot{\vec{r}}_1' + G(m_1+m_2)\vec{r}_1'/d^3 = 0 \\ \ddot{\vec{r}}_2' + G(m_1+m_2)\vec{r}_2'/d^3 = 0 \end{array} \right\} \quad . \tag{6.2.9}$$

We can reduce these further by introducing a new vector that
runs from one object to the other so that

$$\vec{r} = \vec{r}_1' - \vec{r}_2' \quad . \tag{6.2.10}$$

Then by subtracting the second of equations (6.2.9) from the
first we get

$$\ddot{\vec{r}} + GM\vec{r}/d^3 = 0 \quad . \tag{6.2.11}$$

This is equivalent to making another coordinate transformation to one of the objects since $\vec{r}$ is simply the distance between the objects. However, this reduces the problem to the one we solved in the previous chapter, since the form of equation (6.2.11) is the same as equation (5.1.3). Thus the solution of the two body problem is equivalent to the solution of a central force problem where the potential is the gravitational potential and the source of the force can be viewed as being located in one of the objects.

Thus we may jump directly to the solution of the problem given by equations (5.4.9 - 5.4.12) and write

$$\left. \begin{array}{l} r = P/[1\text{-}e\ \cos(\theta\text{-}\theta_0)] \\[2ex] P = L^2/GMm^2 \\[2ex] e = [1+2EL^2/(GMm)^2m]^{\frac{1}{2}} \end{array} \right\} \qquad (6.2.12)$$

Here we have found three more constants in E, L, and θ_0. We knew that the angular momentum and the energy would have to be two of the constants, and that an initial value of θ_0 is involved should be no surprise. While equations (6.2.12) introduce the angular momentum, they only specify its magnitude, and we know from the central force problem that the *vector* is an integral of the motion. That is what insures that the motion is planar. Therefore specifying the angular momentum specifies two additional linearly independent components (in addition to the magnitude). The last remaining constant is the r_0 that appears in equation (5.3.3) and specifies the location of the particle in its orbit at some specific time. Like θ_0 it can be regarded as an initial value of the problem. Thus we have all six remaining constants of the motion containing sufficient information to uniquely determine the position of each object in space as a function of time.

b. Location of the Two Bodies in Space and Time

By choosing a coordinate system with its origin at one of the bodies, we are really only concerned with describing the motion of one of the objects with respect to the other. While equations (6.2.12) indicate the shape of the orbit, they say nothing about how the object moves in time. To describe the motion, we shall have to make use of Kepler's second law, the constancy of the areal velocity. To do this we shall have to introduce some new terminology.

As an example, let us consider the motion of an object about the sun. Since we want to describe the motion of an object in its orbit, we shall need some means to define specific locations in the orbit as reference points and parameters to measure angular positions. We shall presume that the orbit is elliptical with the sun at one focus in accord with Kepler's first law. Thus there will be a point in the orbit where the object makes it closest approach to the sun. This point is known as *perihelion* since, in general, the point of closest approach to the source of the force-field is known as peri*** , where *** is the Greek stem appropriate to the object. This point is always located at one end of the semi-major axis of the ellipse. In the case of orbits about the sun, the other end of the semi-major axis is known as *aphelion* and is the position furthest from the sun. Since the origin of the coordinate system is at the source of the attractive force, the location of the object in its orbit can be defined by an angle measured from the semi-major axis - specifically from the point of perihelion (see Figure 6.1) in the direction of the object's motion. This angle is called the *true anomaly*, and will be denoted by the greek letter ν. Determining it as a function of time essentially solves the problem of finding the temporal location of the object.

Let us choose to start measuring time from perihelion passage so that the true anomaly is zero when t = 0. From the solution to the orbit equation [equation (6.2.12)] we see that t = 0 will occur when $\theta = \theta_0$ so that

$$\nu = \theta - \theta_0 \quad . \tag{6.2.13}$$

We may then write the orbit solution as

$$r = \frac{P}{1+e\ \cos\nu} = \frac{a(1-e^2)}{1+e\ \cos\nu} \quad , \tag{6.2.14}$$

where a is the semi-major axis of the ellipse.

Now we shall appear to digress to some geometry and relate each point on the elliptical orbit to a corresponding point on a circle with a radius equal to the semi-major axis and whose center is located at the *center* of the ellipse (again see Figure 6.1). An ellipse is simply the projection of a circle that has been rotated about its diameter through some angle ψ. Now imagine points $[x_c, y_c]$ located on the circle and corresponding points $[x_e, y_e]$ located on the ellipse. For $x_c = x_e$,

$$\frac{y_e}{y_c} = \frac{b}{a} = \cos\psi \quad , \tag{6.2.15}$$

where a and b are the semi-major and semi-minor axes of the ellipse respectively. Since $\cos\psi$ is the same for all

79

corresponding ($x_c = x_e$) points on the circle and the ellipse, this result must hold for all such points.

The Pythagorean theorem assures us that

$$r^2 = y_e^2 + (f-x_e)^2 = (b/a)^2 y_c^2 + (f-x_c)^2 \quad , \qquad (6.2.16)$$

where f is the distance from the center to the focus of the ellipse. From the equation for the ellipse [see equation (6.2.14)], we can write for $\nu = 0$ that

$$r = a - f = a(1-e^2)/(1+e) = a(1-e) \quad , \qquad (6.2.17)$$

which becomes

$$f = ae \quad . \qquad (6.2.18)$$

If we define an angle E measured from perihelion to a point on the circle [x_c, y_c] as seen from the center of the circle, then

$$\left. \begin{array}{l} x_c = a\ \cos(E) \\[2mm] y_c = a\ \sin(E) \end{array} \right\} \quad . \qquad (6.2.19)$$

Using these definitions and equation (6.2.18), equation (6.2.16) becomes

$$r = a[1-e\ \cos(E)] \quad . \qquad (6.2.20)$$

The angle (E) is called the *eccentric anomaly*. Now we are in a position to relate the areal velocity of the particle along the elliptic orbit to the areal velocity of an imaginary particle along the circle.

Imagine such a particle moving in a circle with a radius equal to the semi-major axis (a) of the ellipse. Both particles would have the same orbital period since that depends only on the semi-major axis. However, the imaginary particle moving on the circle would move along its orbit at a uniform rate of speed. Therefore let us define its angular rate of speed as

$$n = M/t = 2\pi/P \quad , \qquad (6.2.21)$$

where *P* is the orbital period. Here M is the angular distance along the circle that the imaginary particle would have moved during the time t specifying the position of the real particle on the ellipse. Thus

$$M = nt \qquad (6.2.22)$$

The angle M is called the *mean anomaly*.

80

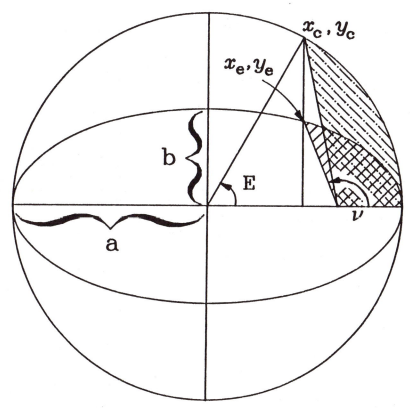

Figure 6.1 shows the geometrical relationships between the elliptic orbit and the osculating circle. The areas swept out by radius vectors to points on the ellipse and the circle are shown as the shaded areas. By relating the sides of the bounded figures, we may relate the area swept out in the ellipse to the area swept out on the circle of a uniformly moving object. This is the source of Kepler's equation.

We may relate the mean anomaly to the eccentric anomaly by the following argument. From the law of areas (Kepler's second law)

$$\frac{M}{2\pi} = \frac{A}{\pi a b} ,$$

(6.2.23)

where A is the area swept out by the radius vector in time t while $\pi a b$ is just the area of the ellipse. Now, since each point on the circle is simply a scaled point on the ellipse, the areas in equation (6.2.23) scale by (a/b) so that

$$\frac{M}{2\pi} = \frac{B}{\pi a^2} = \frac{\frac{1}{2}a^2 E - \frac{1}{2}fy_c}{\pi a^2} = \frac{\frac{1}{2}a^2 E - \frac{1}{2}a^2 e \sin(E)}{\pi a^2} \quad , \tag{6.2.24}$$

where B is the dot-dashed area of Figure 6.1 so that

$$M = E - e \sin(E) \quad . \tag{6.2.25}$$

This expression is known as Kepler's equation since it specifically utilizes Kepler's second law to relate the mean anomaly to the eccentric anomaly. We may use equation (6.2.20) and the equation for an ellipse [equation (6.2.14)] to relate the eccentric anomaly to the true anomaly. By equating the value of r given by each of these equations, we get

$$\frac{a(1-e^2)}{1 + e \cos\nu} = a[1 - e \cos(E)] \quad , \tag{6.2.26}$$

which after some trigonometry becomes

$$\tan(\nu/2) = \left[\frac{1+e}{1-e}\right]^{\frac{1}{2}} \tan(E/2) \quad . \tag{6.2.27}$$

Equation (6.2.27) and Kepler's equation [equation (6.2.25)], therefore, relate the time since perihelion passage to the true anomaly or angular position of the real object in its elliptic orbit.

The conservation of angular momentum leads to similar results for hyperbolic and parabolic orbits. Specifically for hyperbolic orbits we have

$$\left. \begin{aligned}
r &= a[e \cosh(F) - 1] \\[4pt]
M &= e \sinh(F) - F \\[4pt]
\tan(\nu/2) &= [(e+1)/(e-1)]^{\frac{1}{2}} \tanh(F/2)
\end{aligned} \right\} \quad , \tag{6.2.28}$$

while for parabolic orbits we get

$$\left. \begin{aligned}
r &= q \sec^2(\nu/2) = q[1 + \tan^2(\nu/2)] \\[4pt]
2x &= 3M + [9M^2 + 4]^{\frac{1}{2}} \\[4pt]
\tan(\nu/2) &= x^{1/3} - x^{-1/3}
\end{aligned} \right\} \quad . \tag{6.2.29}$$

The quantity n, which is the mean daily motion, has the same physical interpretation for both the elliptic and hyperbolic orbits, but is defined slightly differently for parabolic orbits.

From Newton's laws of motion and gravitation we can write the mean daily motion for objects in elliptic orbit as

$$n = 2\pi/P = [GM/a^3]^{\frac{1}{2}} \quad , \tag{6.2.30}$$

where M is the *sum* of the masses of the two bodies. However, in the solar system we can use the earth's orbital parameters as units to define the motion of objects about the sun and express n in those units and a constant k, known as the Gaussian constant as

$$n = k[(M/M_\odot)/(a/a_\oplus)^3]^{\frac{1}{2}} \quad . \tag{6.2.31}$$

Actually the value of k is taken to be

$$k = 0.01720209895 \quad , \tag{6.2.32}$$

and its value is used to define the astronomical unit. Generally one hears that the astronomical unit is the semi-major axis of the earth's orbit by definition, but this is not strictly correct. It is k that is fixed with units of mass measured in solar masses, time in ephemeris days, and the unit of length is the astronomical unit by definition. Indeed, using the modern value for the mass of the earth (in units of the solar mass) one would find that the semi-major axis of the earth's orbit is about $1+3 \times 10^{-7}$ astronomical units. Brouwer and Clemence[5] point out that Kepler's third law isn't strictly correct if there is a massive third body in the system so the fact that the semi-major axis of the earth's orbit is not exactly one astronomical unit should not be a bother. As long as the unit of length is well defined by equation (6.2.31), we may use it to determine the mean angular motion for objects in the solar system.

The analogous expressions for hyperbolic and parabolic orbits are

$$n = k[(M/M_\odot)/(a_h/a_\oplus)^3]^{\frac{1}{2}} \quad \text{Hyperbolic orbits}$$

$$n = k[(M/M_\odot)/2(q/a_\oplus)^3]^{\frac{1}{2}} \quad \text{Parabolic orbits} \left.\right\} \tag{6.2.33}$$

Here a_h is called the semi-transverse axis of the hyperbola and q is known as the pericentric distance which is simply the distance of closest approach to the second object. In the solar system the sun's mass so dominates that $M/M_\odot$ is effectively unity. Thus if we know the type of orbit and orbital scale

83

length (i.e., semi-major axis for the ellipse, semi-transverse axis for the hyperbola, or pericentric distance for the parabola) we can determine the mean daily motion from equations (6.2.31 - 6.2.33). Further knowledge of the time since perihelion passage allows the calculation of the mean anomaly M. That and the eccentricity enable us to calculate the eccentric anomaly through the solution of Kepler's equation. Algebra, in the form of equations (6.2.27-6.2.29), allows for the calculation of the true anomaly and the radial distance r from the origin of the coordinate system. This, then completely specifies the location of the object in its orbit. Involved as this process is, it is relatively straightforward except for the solution of Kepler's equation.

c. The Solution of Kepler's Equation

Equations of the form of equation (6.2.25) are known as transcendental equations and, in general do not have closed form solutions. Thus, in order to solve the problem of orbital motion, we will be forced to a numerical solution of Kepler's equation. Much has been written on effective and general numerical procedures for such a solution and we will not go into all of those details here. Rather we shall adapt a common numerical procedure known as Newton-Raphson iteration. Assume that we have an equation of the form

$$f(x) = 0 ,$$

$$(6.2.34)$$

and we wish to find that value of x for which the equation is satisfied. A procedure for accomplishing this is to guess an initial value $x^{(0)}$ and use the following expression to improve it.

$$x^{(k+1)} = x^{(k)} - f[x^{(k)}]/f'[x^{(k)}] \quad .$$

$$(6.2.35)$$

The process is repeated until

$$\left| [x^{(k+1)} - x^{(k)}]/x^{(k)} \right| \leq \epsilon \quad ,$$

$$(6.2.36)$$

where ϵ is some predetermined tolerance. The value of x for which $\epsilon = 0$ is known as the 'fixed-point' of the iteration scheme and a rather large body of knowledge has been developed concerning such schemes. The specific one given in equation (6.2.35) has the virtue of normally converging quickly to a fixed point and is simple. It is called Newton-Raphson iteration and graphically amounts to extending a tangent to the function $f[x^{(k)}]$ to the point where it intercepts the x-axis and using that value of x as $x^{(k+1)}$. Clearly, when f(x) is

84

zero, x is a fixed point. The application of the method to Kepler's equation yields

$$\left.\begin{array}{l} E^{(0)} = M + e \sin(M) \\ E^{(k+1)} = E^{(k)} + [M - E^{(k)} + e \sin(E^{(k)})]/[1 - e \cos(E^{(k)})] \end{array}\right\} \quad (6.2.37)$$

One of the problems with the Newton-Raphson scheme is that it doesn't always converge. This is the case with equations (6.2.37). There are values of the eccentricity and mean anomaly for which this iteration scheme will not yield an answer. However, this occurs only for a small range of M near perihelion and very large eccentricities (see chapter 6 exercises). It will always work for objects in elliptical orbits in the solar system except for some long period comets and these orbits may be handled in another manner. Thus, for simplicity, we will leave the discussion of the solution of Kepler's equation with the Newton-Raphson iteration scheme. Those who wish more details on the subject should consult Green[6].

6.3 The Orientation of the Orbit and the Orbital Elements

The solution to the two body problem consists in describing the motion of both bodies in an arbitrary coordinate frame. Since the two bodies are described by two vector differential equations of second order, there will be twelve constants required for that description. Six of those twelve are required to describe the motion of the center of mass of the system. Three more are required to locate one object in its orbit relative to the other. The remaining three are required to specify the orientation of the orbit with respect to the arbitrary coordinate frame. If we assume that the coordinate frame is a spherical coordinate frame, then we can use the Euler angles as defined in chapter 2 to define the orbital orientation in that frame. The coordinate frame will have a fundamental plane and a direction within that plane that defines how azimuthal angles will be measured. For most astronomical coordinate systems of relevance to celestial mechanics, that direction is toward the first point of Aries (i.e., the vernal equinox) and the fundamental plane will be either the ecliptic or the equator of the earth (see chapter 2).

Figure 6.2 shows the orbit of an object located in the reference coordinate frame and it bears a marked similarity to the last of Figures 2.2. In Figure 2.2 ϕ described the distance from the preferred direction to the line of intersection of the two planes known as the line of nodes. In celestial mechanics, this is known as the longitude of the ascending node where the notion of "ascending" refers to that node where the motion of the object carries it toward positive Z. In the solar system, this means that the object would be moving from south to north in the sky. We will use Ω to denote this angle. The second of the Euler angles in Figure 2.2 is θ and measures the angle by which one plane is inclined to the other. In celestial mechanics this is known as the angle of inclination and is usually denoted by i. The last of the Euler angles in Figure 2.2 is ψ and is used to denote a particular point in the inclined plane. For orbital mechanics the most logical point in the orbit is the pericenter. Its location is then designated by the angle ω called the argument of the pericenter. Thus the three defining angles of the orbit are

$\Omega \equiv$ The Longitude of the Ascending Node

$i \equiv$ The Inclination of the Orbit
 (measured from $0° \rightarrow 180°$)

$\omega \equiv$ The Argument of the Pericenter
 (measured from the ascending node in the
 direction of motion with a range of $0° \rightarrow 360°$)

$$(6.3.1)$$

Sometimes the argument of the pericenter is replaced by the strange angular sum $\omega + \Omega$ which is called the longitude of the pericenter and is denoted by

$$\tilde{\omega} \equiv \Omega + \omega \equiv \text{The Longitude of the Pericenter} \qquad . \qquad (6.3.2)$$

Thus we have defined the three remaining constants required by the equations of motion specifying the orientation of the orbital plane. In the solar system, the center of attraction is usually the sun and so the pericenter becomes perihelion and the fundamental plane is usually the ecliptic.

We have repeatedly said that there are twelve constants required to uniquely specify the motion of one object about another, but that six of them are concerned with the motion of the center of mass of the pair. Since this motion is uniform, these six constants are usually ignored when discussing the orbit of the object. The remaining six constants constitute the elements of the orbit and can be broken into two sets of three. The three that define the orientation of the orbit as defined above are taken directly as orbital elements. However, the remaining three that specify the size and shape of the orbit as well as the object's location in it at some time can be specified in various ways. We found in chapter 4 that the angular momentum and total energy are integrals of the motion and will determine the size and shape of the orbit. However, they are not directly observable quantities so that a different set of constants more directly related to the geometry of the orbit is usually chosen to represent the orbit. These are the semi-major axis and the eccentricity. Finally to represent the position of the object within its orbit we specify the time when the object is at pericenter, or for the solar system, the time of perihelion passage T_0. Now in developing the equations describing the motion of the object in its orbit, we took the time of perihelion passage to be zero. Thus (t) in equation (6.2.21) and equation (6.2.22) should be replaced by

$$t = t-T_0 \quad . \tag{6.3.3}$$

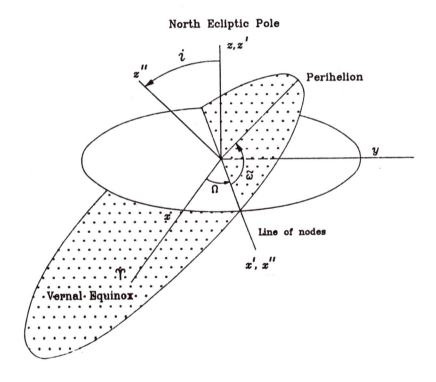

Figure 6.2 shows the coordinate frames that serve to define the orbital elements specifying the orientation of the orbit with respect to the ecliptic coordinate system.

87

The six constants specifying the motion of the object are known as the *elements of the orbit* of the object and are

a ≡ The Semi-major Axis of the orbit

e ≡ The Orbital Eccentricity

T_0 ≡ The Time of Perihelion Passage

ω ≡ The Argument of Perihelion

Ω ≡ The Longitude of the Ascending Node

i ≡ The Inclination of the Orbit

(6.3.4)

While we have now located the object in its orbit, we have yet to find it in the sky.

6.4 The Location of the Object in the Sky

The location of the object in the sky involves nothing more than the transformation from the coordinate system specifying the location of the object in its orbit to the coordinate system of the observer. The specific nature of this transformation depends on the relative location of the source of the attractive force and observer. For example, we will consider the object to be in orbit about the sun and the observer located on a spinning earth. Since the heliocentric orbital elements are generally referred to the ecliptic, the first part of the transformation will involve expressing the components of the radius vector to the object in ecliptic coordinates. Then we will transform to the equatorial (Right Ascension-Declination) coordinate system. This is followed by shifting the origin of the coordinate system to the center of the earth and finally the astronomical triangle may be solved to express the result in the observer's Alt-Azimuth coordinate system.

Imagine a cartesian coordinate system with its origin coinciding with the sun, the z-axis normal to the orbit plane, and the x-axis passing through perihelion. In such a coordinate system the components of the radius vector to the orbiting object are

$$\vec{r} = \begin{bmatrix} r \cos\nu \\ r \sin\nu \end{bmatrix}$$

(6.4.1)

We wish to transform this coordinate frame to the equatorial coordinate frame. Therefore we first carry out the *inverse* Euler rotational transformations that will align the x-axis with the direction to the vernal equinox and the z-axis normal to the plane defining the orbital elements (usually the ecliptic plane). This will yield the components of the vector in ecliptic coordinates as

$$\vec{r}' = P_z^T(\Omega) P_x^T(i) P_z^T(\omega) \vec{r} \qquad . \qquad (6.4.2)$$

Now to express the coordinates in Right Ascension-Declination coordinates, we must align the defining planes of the two coordinate systems. This can be accomplished by a rotation about the x-axis, pointing toward the vernal equinox, through an angle $-\epsilon$ where ϵ is the angle between the ecliptic and equatorial planes. Note that a rotation through a negative angle is equivalent to the inverse transformation of the positive rotation. Thus the radius vector can be expressed in heliocentric equatorial coordinates as

$$\vec{r}'' = P_x^T(\epsilon) \vec{r}' \qquad . \qquad (6.4.3)$$

Now the origin of the coordinate system must be transferred to the earth. This is a vector transformation and is accomplished by simply subtracting a heliocentric vector to the earth from the heliocentric vector locating the object. Thus a radius vector from the *earth* to the object will have geocentric equatorial coordinates of

$$\vec{\rho} = [P_x^T(\epsilon) P_z^T(\Omega) P_x^T(i) P_z^T(\omega)] \vec{r} - \vec{X}_\oplus \qquad . \qquad (6.4.4)$$

Here the vector $\vec{X}_\oplus$ is the heliocentric equatorial radius vector to the earth.

Having arrived at the earth, we need only correct for the observer's location on the earth. Remember that the x-axis is still pointing at the vernal equinox and the z-axis toward the north celestial pole. Thus to get to the local alt-azimuth coordinate system, we must align the x-axis with the local prime meridian (pointing north) and then bring the z-axis so that it points toward the zenith. The first of these transformations can be accomplished by rotating about the z-axis (polar axis) through the local hour angle of the vernal equinox, but this is just the local sidereal time by definition. At this point the x-axis will lie in the plane of the prime meridian, but pointing south (in the northern hemisphere) so we must rotate through an additional angle of 180°. If the object happens to be close by, it may finally be necessary to transfer the origin from the center of the earth to the observer by subtracting the radius vector from the

89

center of the earth to the observer's location. Following this by a rotation through the co-latitude of the observer will bring the z-axis so that it points toward the zenith. Thus the complete transformation from the orbital coordinates of equation (6.4.1) to the true topocentric coordinates of the observer can be written as

$$\vec{\rho}' = P_y(\pi/2-\delta)P_z[h(t)]\{P_x^T(\epsilon)P_z^T(\Omega)P_x^T(i)P_z^T(\omega)\vec{r} - \vec{X}_\oplus\} - \vec{r}_\oplus. \quad (6.4.5)$$

If the transformation from the center of the earth to the true topocentric coordinates is carried out as indicated by equation (6.4.5), the vector $\vec{r}_\oplus$ has only an x-component equal to the radius of the earth for the observer's latitude and longitude. The components of the vector from the observer have the following components in the Alt-Azimuth coordinate system :

$$\vec{\rho} = \begin{pmatrix} \rho\sin(H) \\ \rho\cos(H)\cos(A) \\ \rho\cos(H)\sin(A) \end{pmatrix}, \quad (6.4.6)$$

which translates into the Alt-Azimuth coordinates of

$$\left. \begin{aligned} \rho^2 &= \rho_x^2 + \rho_y^2 + \rho_z^2 \\ \tan A &= \rho_y/\rho_x \\ \sin H &= \rho_x/\rho \end{aligned} \right\} \quad (6.4.7)$$

Thus we have completely described the motion of a object around the sun to the point where we can locate the object in the sky. In the next chapter we shall consider the inverse problem of determining the orbital elements from observation.

Chapter 6: Exercises

1. Given a body which is bounded by the surface

    $$x^2(b^2+3a^2) + 2\sqrt{3}(xy)(b^2-a^2) + y^2(3b^2+a^2)$$

    $$+ (4a^2b^2/c^2)z^2 = 4a^2b^2 ,$$

    where $a > b > c$, and has a density distribution $\rho(r) = $ const, find the principal moments of inertia and the principal axes of the body.

2. Integrate the equations of motion for the two-body problem to show that

    $$\epsilon = (1+2EL^2/mk^2)^{\frac{1}{2}} .$$

3. Assuming the earth's orbit to be circular and that meteors approach the sun in parabolic orbits, between what limits on their relative speed will they hit the earth if the gravitational attraction of the earth is neglected?

4. Consider two particles orbiting about one another and having masses m_1 and m_2. If the force between the two is given by

    $$\vec{F} = k^2(\vec{r}_1 - \vec{r}_2) ,$$

    show that the orbit of one particle about the other is an ellipse with one particle at the *center* of the ellipse.

5. A rocket is detected approaching Chicago at a range of 3200km, and an altitude of 160km above sea level. If the velocity of approach is 24800km/hr and the motion is parallel to the surface of the earth, decide if the rocket will hit Chicago. Assume that the earth is spherical and that coriolis forces and atmospheric drag are neglible. What are the values of r and ν at the instant of detection? If it should miss, how much will it miss by? If the azimuth at the time of detection is 15°, where is the probable launch site?

6. Find the Right Ascension, declination, altitude and azimuth for Mars as seen from The Ohio State University campus on March 1, 1988 at 3:00AM EST. List all additional constants and their source necessary to solve this problem.

7.　　If one has an iterative function that can be written as

$$x^{(k+1)} = \Im[f(x^{(k)})] \ ,$$

then it will converge to a fixed point if and only if

$$\left|\frac{\partial \Im}{\partial x}\right| < 1 \text{ for all x such that}$$

$$|x^{(k)}| < |x| < |x_{(\text{fixed-point})}| \qquad .$$

Find the range of values of e and E for which Newton-Raphson iteration will converge to a solution of Kepler's equation.

7

The Observational Determination of Orbits

In the last chapter we saw how to find the position on the sky of a object given the parameters that describe the orbit of the object. That is about half of the fundamental problem of celestial mechanics. The other half is the reverse. Namely, given some observational information about the motion of the object, one would like to determine the orbital elements that specify the motion. This, and chapter 6, enable one to predict the future location and motion of the object. These two parts of the description of orbital motion constitute the solution of the primary problem of celestial mechanics.

It is clear from what we have done in chapters 4 and 5 that the solution of the equations of motion for n-bodies requires 6n constants of integration. For two bodies half of the constants are involved in describing the motion of the center of mass, while the remaining six specify the location of a particle in its orbit and the orientation of that orbit with respect to a specified coordinate system. Thus, for objects in orbit about the sun we have only the six orbital elements that represent the six linearly independent constants required for the solution of the equations of motion. In order to determine these six linearly independent orbital elements, we will need six linearly independent pieces of information. There are many different forms that this information may have. For example, one might have the position and velocity at some instant in time. These two vectors clearly provide six independent pieces of information as they constitute the classical initial values for the integration of the Newtonian equations of motion. However, they are not the quantities traditionally available to the astronomer. Classically, one observes the position of an object as seen projected against the celestial sphere. Such an observation is comprised of two angular coordinates and the time of observation. This represents two linearly independent pieces of information so that one would need three such observations in order to determine the orbital elements. In principle, one might also be able to measure the radial velocity with respect to the earth, but this is only one additional independent piece of information. Thus, if one had two positions and the radial velocities of the object at those positions, the problem would be determined.

In practice, all observations are subject to error and this will be reflected in errors in the orbital elements. Therefore, the accurate determination of orbital elements will make use of a large number of observations combined in such a way as to reduce the resultant error of the final result. The combination of the observations usually employs some principle such as Legendre's principle of least squares or more contemporarily, the related maximum likelihood principle. However, all of these methods require the relationship between the orbital elements to be determined and the particular type of observations to be specified. Since this relationship is, in general, nonlinear, we shall consider several different and specific cases. As an example and for traditional reasons, we shall consider the problem of determining the orbital elements for an object in orbit about the sun. However, the approaches are much more general and are applicable for determining the orbits of objects revolving about most any object where the potential is that of a point mass.

7.1 Newtonian Initial Conditions

In chapters 3 and 5 we found that the two body problem will have two integrals of the motion, the angular momentum and the total energy. Integrals of the motion are useful for our purpose since they are indeed constant for all parts of the orbit and therefore apply as constants for all possible observations. They represent constraints that all observations must satisfy, and they can be directly related to the orbital elements. Therefore we will begin by discussing what they can tell us about positions and velocities and vice versa.

Let us assume that we know a position and velocity at some instant in time. This is essentially the initial value information that would be needed for the direct solution of the Newtonian equations of motion. The definition of angular momentum requires that

$$\vec{r} \times \dot{\vec{r}} = \vec{L}/m = |r||\dot{r}|\sin\theta \quad , \tag{7.1.1}$$

and the angular momentum is an integral of the motion. From the solution of the two body problem [see equations (6.2.12), and (6.2.14)] and the properties of an ellipse we know that

$$P = L^2/GMm = a(1-e^2) \quad .$$ (7.1.2)

If we combine this with the expression for the velocity of an object moving in a central force field [equation (5.4.11)], we get

$$\dot{\vec{r}} \cdot \dot{\vec{r}} = v^2 = GM \left[\frac{2}{r} - \frac{1}{a} \right] \quad .$$ (7.1.3)

This is often called the energy integral since it is basically an expression for the conservation of energy. Older books on celestial mechanics refer to it by its old Latin name of the *vis viva* Integral. It immediately supplies us with a value for the semi-major axis

$$a = GMr/(2GM-rv^2) \quad ,$$ (7.1.4)

which is one of the desired orbital elements.

Now we may obtain an expression for the true anomaly by solving equation (6.2.14) for (e cosν), differentiating it with respect to time, and remembering that $\dot{\nu}$ is related to the areal velocity which is a integral of the motion [see equation (5.3.1)], so that

$$\frac{P}{r^2} \dot{r} = e(\sin\nu)\dot{\nu} = \frac{eL}{mr^2} \sin\nu \quad .$$ (7.1.5)

When we eliminate P by using equation (7.1.2) and L with the aid of equation (7.1.1) we get

$$\sin\nu = L\dot{r}/GMm = \dot{r}^2 r(\sin\theta)/GM \quad .$$ (7.1.6)

Remember that sinθ is known by virtue of knowing $\vec{r}$ and $\dot{\vec{r}}$. We may now use the solution for the elliptic orbit [equation (6.2.14)] to write

$$1+e \cos\nu = [|\dot{r}|\sin\theta]GM \quad .$$ (7.1.7)

This provides a value for the orbital eccentricity e.

The true anomaly and the eccentricity allow us to directly calculate the eccentric anomaly (E) from equation (6.2.27) and, by means of Kepler's equation [equation (6.2.25)], the mean anomaly M. The mean anomaly, in turn, allows us to calculate the time of perihelion passage since the mean daily motion (n) depends only on the period which, in

95

turn, depends only on the semi-major axis so that

$$T_0 = t_1 - [E - e \sin(E)]/n = t_1 - (a/GM)^{\frac{1}{2}}a[E - e \sin(E)]. \quad (7.1.8)$$

Here t_1 refers to the time at which the observations of the position and velocity are made. Thus equations (7.1.4), (7.1.7), and (7.1.8) determine the shape of the orbit and the orbital element that locates the object in its orbit. The information that has been used to determine these orbital elements is just the magnitude of the angular momentum and energy and the angle between the position and velocity vector. These are three linearly independent pieces of information and they determine three orbital elements. Clearly the energy and angular momentum determine the shape and size of the orbit as they are integrals of the motion and are constants for all points in the orbit. Taken together with the angle between the position and velocity vectors, they are sufficient to locate the particle in that orbit.

The remaining three orbital elements specify the orientation of the orbit and must be determined from information uniquely related to its orientation. The angular momentum vector always points normal to the orbit and, being an integral of the motion, is sufficient to specify the orbit's orientation. A unit vector pointing in the direction of the angular momentum vector contains all the information necessary to specify the orbital orientation. It can be specified in terms of the position and velocity vectors as

$$\hat{n} = \vec{L}/L = [\vec{r} \times \dot{\vec{r}}] / |r||\dot{r}| \sin\theta \quad . \quad (7.1.9)$$

Thus, the components of that vector in any particular coordinate system will specify the orientation of the orbit in that coordinate system. The components in the ecliptic coordinate system yield two of the remaining three orbital elements from

$$
\left.
\begin{aligned}
n_x &= \sin\Omega \, \sin(i) \\
n_y &= -\cos\Omega \, \sin(i) \\
n_z &= \cos(i)
\end{aligned}
\right\} \qquad (7.1.10)
$$

The remaining orbital element can be determined by considering a unit vector ($\hat{\eta}$) pointing toward the vernal equinox and its scalar and vector products with the position vector $\vec{r}$ which are

$$
\left.
\begin{aligned}
\hat{\eta} \cdot \vec{r} &= r \cos(\nu + \omega) = r_x \cos\Omega + r_y \sin\Omega \\
\hat{\eta} \times \vec{r} &= r \sin(\nu + \omega)\hat{n}
\end{aligned}
\right\} \qquad (7.1.11)
$$

The x-component of the vector cross product is

$$(\hat{\eta} \times \vec{r})_x = r_z \sin\Omega = r \sin\Omega \sin(i) \sin(\nu+\omega) \quad , \qquad (7.1.12)$$

which along with the scalar product yields

$$\left. \begin{aligned} \sin(\nu+\omega) &= r_z / [r \sin(i)] \\ \cos(\nu+\omega) &= r_x \cos\Omega + r_y \sin\Omega \end{aligned} \right\} \qquad (7.1.13)$$

These two equations are sufficient to unambiguously determine $(\nu+\omega)$ and hence the last remaining orbital element, the argument of perihelion ω.

Thus we have seen how, given what amount to initial conditions of the motion, $[\vec{v}(t_1), \vec{r}(t_1)]$ can be used to determine the orbital elements. It is important to recognize the type of information available and which orbital elements are constrained by that information. Magnitudes of position and velocity vectors specify the magnitudes of the orbital energy and angular momentum. Since these are integrals of the motion, they will determine the size and shape of the orbit. The constancy of the angular momentum vector in space will essentially determine the orientation of the orbit. A combination of both is required to locate the object is its orbit. All methods of determining orbital elements will utilize the observed information in this way. While astronomers rarely are able to determine position and velocity vectors at a given instant, most methods of orbit determination rely on estimating this information from the information that is available.

7.2 Determination of Orbital Parameters from Angular Positions Alone

The traditional problem of celestial mechanics involves the determination of the orbital elements given the angular position on the celestial sphere at various times. This information takes the form of pairs of celestial coordinates in some known coordinate system. Since there are six constants of the motion, we will need at least six independent observational constraints or three observations of coordinate pairs. To understand conceptually how this can work, let us consider a method that dates back at least to Johannes Kepler.

a. The Geometrical Method of Kepler

This method determines the planetary orbit with respect to the earth's orbit. In a way this is true for all methods

97

since the scale of the solar system is set by the value of the astronomical unit which is generally assumed to be known. However, it is interesting that this method makes no use of physics and only assumes that both the earth and planet are in orbit about the sun. Indeed, this is the method by which Kepler discovered his laws of motion. One begins by determining the sidereal period of the planet, the time required for the planet to return to the same point with respect to the stars as seen in an inertial frame. This is done by measuring the synodic period directly. The synodic period is simply the length of time required for the planet to return to the same place in the sky *as seen from the earth* (see Figure 7.1).

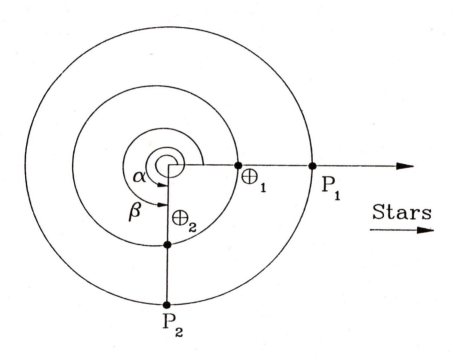

Figure 7.1 shows the orbital motion of a planet and the earth moving from an initial position with respect to the sun (opposition) to a position that repeats the initial alignment. This associated time interval is known as the synodic period of planet P with respect to the earth. The concept of a synodic period need not be limited to the earth and another planet, but may involve any two planets.

Let this period of time be P_s. Now the angular distance traveled by the earth during this time will just be $(2\pi/P_\oplus)P_s$ where $P_\oplus$ is the sidereal period of the earth. During the same interval of time the planet will have traveled an angular distance $(2\pi/P_p)P_s$. However, since the planets have returned to the same relative position in the sky with respect to the sun, the angular difference in the distance traveled must be 2π. Therefore

$$\left| \frac{2\pi}{P_\oplus} - \frac{2\pi}{P_p} \right| = \frac{2\pi}{P_s} \qquad . \qquad\qquad (7.2.1)$$

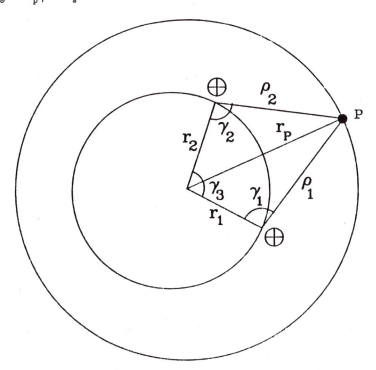

Figure 7.2 shows the position of the earth at the beginning and end of one sidereal period of planet P. If we assume that the distance of the earth to the sun as well as the three angles γ_i, are known at each position, then the determination of the remaining parts of the quadrilateral, including the distance to the planet, is a matter of plane trigonometry.

99

Thus careful observation of the synodic period will lead to the determination of the sidereal period of the planet. While it is true that elliptic orbits will cause difficulties with this approach, it is possible to wait a number of synodic periods until the planet returns arbitrarily close to a given position in the sky and then the method will give the correct result in spite of the orbital eccentricity. While this is not strictly an angular position, it is the measurement of a single item, the synodic period, which then specifies the sidereal period.

Now simply observe the position of the planet at the beginning and end of one *sidereal* period. It will be seen against the stellar field from two different vantage points as the sidereal period of the planet will not in general be commensurate with that of the earth. Thus the planet will lie at the vertex of a quadrilateral formed by the planet, sun and two positions of the earth (see Figure 7.2). Assuming that the orbit of the earth is known, then two sides and three angles of the quadrilateral are known. This enables the remaining sides and diagonals to be determined. If this procedure is carried out throughout the entire orbit of the planet, its entire orbit with respect to the earth can be measured. If the detailed shape of the orbit is known, then clearly the orbital elements that describe the orbit are specified. Much more than the minimum three pairs of observations have gone into this determination, but much less has been assumed. The two-body orbital mechanics that gives rise to the six constants of motion and even allows us to say what minimum amount of information is necessary has not even been used. Let us now consider a method that integrates Newtonian mechanics into the geometrical approach of Kepler.

b. The Method of Laplace

The basic approach of Laplace was to write the equations of motion in terms of the change of a vector from the earth to the object and then to separate the vector into its magnitude and its direction cosines. It is the changes in these direction cosines that essentially constitute the angular measurements that determine the orbital elements. The entire procedure estimates the values for the position and velocity vectors at some instant in time. One then can use the procedure in section 7.1 to get the orbital elements. This is the schematic procedure that we will follow, but to begin we shall make the following definitions for the vectors involved in the development:

100

$\vec{r}_p$ = the heliocentric radius vector to the planet

$\vec{r}_\oplus$ = the heliocentric radius vector to the earth   (7.2.2)

$\vec{\rho}$ = a vector from the earth to the planet

Now let us represent the components of the vector $\vec{\rho}$ from the earth to the object by their direction cosines specified in terms of the equatorial coordinates of the object so that

$$\vec{\rho} = \rho \begin{pmatrix} \cos\delta \; \cos\alpha \\ \cos\delta \; \sin\alpha \\ \sin\delta \end{pmatrix} = \rho\vec{\lambda} = \rho\hat{\rho} \qquad . \qquad (7.2.3)$$

The vector $\vec{\lambda}$ is simply a unit vector pointing from the earth to the object. I have deliberately continued to use the older notation for the geocentric distance ρ rather than the currently accepted symbol Δ as the latter has too widely an accepted interpretation as the finite difference operator.

    The radial equations of motion for both the object and the earth are

$$\left. \begin{array}{l} \dfrac{d^2\vec{r}_p}{dt^2} = -\dfrac{k\vec{r}_p}{r_p^3} \\[3mm] \dfrac{d^2\vec{r}_\oplus}{dt^2} = -\dfrac{k\vec{r}_\oplus}{r_\oplus^3} \end{array} \right\} \qquad (7.2.4)$$

where

$$\vec{r}_p = \vec{\rho} + \vec{r}_\oplus \qquad . \qquad (7.2.5)$$

If we use this to eliminate $\vec{r}_p$ from the first equation of motion we get

$$\frac{d^2\vec{\rho}}{dt^2} + \frac{d^2\vec{r}_\oplus}{dt^2} = \frac{d^2\vec{\rho}}{dt^2} - \frac{k\vec{r}_\oplus}{r_\oplus^3} = -\frac{k}{r_p^3}[\vec{\rho}+\vec{r}_\oplus] \qquad , \qquad (7.2.6)$$

and using this result we can eliminate the earth's acceleration from its equation of motion and arrive at

$$\frac{d^2\vec{\rho}}{dt^2} + \frac{k\vec{\rho}}{r_p^3} = k\vec{r}_\oplus \left[\frac{1}{r_\oplus^3} - \frac{1}{r_p^3} \right] \qquad . \qquad (7.2.7)$$

101

Explicitly differentiating the vector $\vec{\rho}$ we get

$$\frac{d^2\vec{\rho}}{dt^2} = \frac{d^2(\rho\vec{\lambda})}{dt^2} = \ddot{\rho}\vec{\lambda} + 2\dot{\rho}\dot{\vec{\lambda}} + \rho\ddot{\vec{\lambda}} \quad , \qquad (7.2.8)$$

which when substituted into equation (7.2.7) yields

$$\ddot{\rho}\vec{\lambda} + 2\dot{\rho}\dot{\vec{\lambda}} + \rho\ddot{\vec{\lambda}} = -\frac{k\vec{\rho}}{r_p^3} + k\vec{r}_\oplus \left[\frac{1}{r_\oplus^3} - \frac{1}{r_p^3} \right] \qquad (7.2.9)$$

Regrouping the terms so that the time derivatives of ρ are collected we get

$$\ddot{\rho}\vec{\lambda} + 2\dot{\rho}\dot{\vec{\lambda}} + \rho\left[\ddot{\vec{\lambda}} + \frac{k\vec{\lambda}}{r_p^3} \right] = k\vec{r}_\oplus \left[\frac{1}{r_\oplus^3} - \frac{1}{r_p^3} \right] \qquad (7.2.10)$$

Except for these time derivatives and $\vec{r}_p$, all the parameters of equation (7.2.10) are known. Remember this is a vector equation so that it constitutes three scalar equations for $\ddot{\rho}$, $\dot{\rho}$, and ρ. The right hand side involves k and the heliocentric radius vector to the earth $\vec{r}_\oplus$, which is presumed to be known for all the times of the observations. The parameter r_p may be expressed in terms of ρ, $r_\oplus$, and the angle ψ from the law of cosines as

$$r_p^2 = \rho^2 + r_\oplus^2 - 2\rho r_\oplus \cos\psi \quad . \qquad (7.2.11)$$

However, this angle can be obtained from the scalar product of $\vec{r}_p$ and $\vec{\lambda}$ as

$$\cos\psi = (\vec{r}_\oplus \cdot \hat{\rho})/|r_\oplus| = (\vec{r}_\oplus \cdot \vec{\lambda})/|r_\oplus| \quad . \qquad (7.1.12)$$

Thus, equations (7.2.10-7.2.12) form a closed system of equations for ρ and its first two time derivatives. This must be solved numerically and by iteration due to the nonlinearity of equations (7.2.11), and (7.2.12). Of course the solution depends on having values of $\vec{\lambda}$ and its time derivatives.

For these time derivatives we turn to the observations. Each positional observation consists of a pair of angular coordinates (α, δ) at some particular time t_i. These angular coordinates are sufficient to generate all the components of $\vec{\lambda}$ from equation (7.2.3). Thus three temporal measurements provide three values of the vector $\vec{\lambda}$. Now expand this vector in a Taylor series in time about the first observation so that

102

$$\vec{\lambda}(t) = \vec{\lambda}(0) + \dot{\vec{\lambda}}(0)t + \tfrac{1}{2}\ddot{\vec{\lambda}}(0)t^2 + \cdots + \qquad . \qquad (7.2.13)$$

Thus, for the three successive times of observations we can write

$$\vec{\lambda}(t_i) = \vec{\lambda}(0) + \dot{\vec{\lambda}}(0)t_i + \tfrac{1}{2}\ddot{\vec{\lambda}}(0)t_i^2 + \cdots +, \quad i = 1\cdots3 . \quad (7.2.14)$$

These constitute three linear algebraic vector equations in $\vec{\lambda}$ and its time derivatives. Their solution need only be done once per problem as they provide the constants necessary for the iterative solution of equations (7.2.10-7.2.12) for ρ and its time derivatives. However, the solution of these equations is where most of the error in the final solution arises. If the observations are taken too close together, then their linear independence becomes weak and their values (particularly for the second time derivative) small to indeterminate. Simply, too small a section of the orbit is sampled to provide an accurate determination of the orbital elements. If they are taken too far apart in time, then the validity of the Taylor series becomes suspect. In practice, one would use a number of observations and perhaps a longer Taylor series to ensure that the first three terms were accurately determined. Having assured the accurate determination of $\vec{\lambda}$ and its derivatives, one can turn to the solution of equations (7.2.10-7.2.12) and obtain values for ρ and its time derivatives. These and $\vec{\lambda}$ determine the position vector for the object in heliocentric coordinates and its time derivative yields the velocity vector for the object, all at the time of the first observation so that

$$\left. \begin{aligned} \vec{r}_p &= \rho\vec{\lambda} + \vec{r}_\oplus \\ \dot{\vec{r}}_p &= \vec{v}_p = \rho\dot{\vec{\lambda}} + \dot{\rho}\vec{\lambda} + \dot{\vec{r}}_\oplus \end{aligned} \right\} \qquad (7.2.15)$$

We may now use the methods described in the previous section to find the actual orbital elements. Let us turn to a rather more elegant method that avoids many of the problems of the method of Laplace.

c. The Method of Gauss

While the method of determining orbital elements devised by Laplace is conceptually straightforward, it tends to produce poor initial orbital elements. The reason for this lies in the approximation for the temporal behavior of the radius vector $\vec{\rho}$ from the earth to the object. The Taylor series approximation used to obtain derivatives of $\vec{\rho}$ will generally give uncertain values for those derivatives, which, because of the nonlinearity of the problem, yield poor values for the orbital elements. Another approach to the problem, due to Gauss, while more complicated, usually produces more accurate results. The reason is that the method of Gauss makes approximations to the dynamics of the motion but treats the geometry of the observations in a precise manner. The error propagation of this approach is generally less unstable than that of the method of Laplace. However, due to the detailed complexity of the method, we will only review the conceptual approach here and refer the student to Danby[7] or Moulton[8] for the details.

Gauss begins by taking advantage of the fact that motion of any object about the sun (or any two body problem) takes place in a plane. Thus it is possible to represent the radius vector from the sun to the object in question for any of the three observations as a linear combination of the other two so that

$$\vec{r}_{pi} = C_j \vec{r}_{pj} + C_k \vec{r}_{pk} \quad , \quad i \neq j \neq k; \ i = 1,2,3 \quad . \qquad (7.2.16)$$

These represent three vector equations for the values of $\vec{r}_p$, but they are not linearly independent. However, if we introduce the fact that the observations are made from a moving platform (i.e., the earth) by making use of equation (7.2.5), we can generate three vector equations for the geocentric radius vector of the object $\vec{\rho}_i$ and these are linearly independent. These vector equations are

$$\vec{\rho}_i - C_j \vec{\rho}_j - C_k \vec{\rho}_k = C_j \vec{r}_{\oplus j} + C_k \vec{r}_{\oplus k} - \vec{r}_{\oplus i} \quad , \quad i \neq j \neq k \ ;$$

$$i = 1,2,3 \quad . \qquad (7.2.17)$$

If the C_js which determine that fraction of each vector required to produce the third were known then everything on the righthand side of equations (7.2.17) would be known and we could solve for three values of the geocentric radius vectors $\vec{\rho}_i$. Remember that only the magnitude of $\vec{\rho}_i$ is unknown as the direction cosines are the observations as given in equation (7.2.3). With those three values and the three heliocentric radius vectors of the earth $\vec{r}_{\oplus i}$ we can calculate three values for the heliocentric radius vector of the object $\vec{r}_{pi}$. Given three values for the heliocentric radius vector, there are a number of ways to proceed to obtain the orbital elements. It would appear that there is more information here than is necessary as the three heliocentric radius vectors have nine independent components where only six are required. However, only two of the radius vectors can be regarded as being truly linearly independent. But that is enough. Gauss himself gave a complicated method involving Kepler's equation for obtaining the elements from the three heliocentric radius vectors. Others have used the three heliocentric radius vectors to generate $\dot{\vec{r}}_{p2}$ which, when coupled with $\vec{r}_{p2}$ reduces the problem to the initial value problem that we discussed in section 7.1. Thus all that remains is to find an expression for the C_js.

Consider taking the vector cross product of equation (7.2.16) with $\vec{r}_{pj}$ to get

$$\vec{r}_{pi} \times \vec{r}_{pj} = r_{pi} r_{pj} \sin\theta_{ij} \hat{\ell} = C_k \vec{r}_{pk} \times \vec{r}_{pj}$$

$$= C_k r_{pk} r_{pj} \sin\theta_{kj} \hat{\ell} \ , \quad i \neq j \neq k \ ; \qquad i = 1,2,3 \ . \qquad (7.2.18)$$

The vector $\hat{\ell}$ points normal to the orbit and could be used to determine the orbital elements associated with the orientation of the orbit once the $\vec{r}_{pj}$'s are known. The scalar coefficients of $\hat{\ell}$ are just the areas of the triangles formed by $\vec{r}_{pi}$ and $\vec{r}_{pj}$ (see Figure 7.3). Thus, it is clear that the C_ks are given by

$$C_k = (r_{pi} r_{pj} \sin\theta_{ij})/(r_{pk} r_{pj} \sin\theta_{kj})$$

$$= A_{ij}/A_{kj} \ , \quad i \neq j \neq k \ ; \qquad i = 1,2,3 \ , \qquad (7.2.19)$$

where A_{ij} is the area of the triangle $SP_i P_j$. If the area of the triangle were the area of the orbital sector enclosed by $\vec{r}_{pi}$ and $\vec{r}_{pj}$ then Kepler's second law would guarantee that C_k would simply be given by the ratio of the appropriate time interval between observations so that

$$C_k \approx |t_i - t_j|/|t_k - t_j| \ . \qquad (7.2.20)$$

Here the linear dependence of the three heliocentric radius vectors is clearly displayed as $C_2 = [C_3/C_1]$. However, since we only need two heliocentric radii to solve the problem, we may reduce the number of equations in (7.2.17) to just two which will then be linearly independent.

The complicated part of the method of Gauss is involved in calculating corrections to the triangular area so that it will approximate the sector. Since the corrections appear both in A_{ij} and A_{kj} they will tend to cancel to first order and so need not be terribly accurate. This clearly demonstrates the cleverness of Gauss and the reason for the superiority of his method to that of Laplace. The truncation errors of the Taylor series for the time derivatives of ρ enter directly into the determination of the orbital elements. However, the approximation of Gauss is in the geometric representation of the orbital motion of the object and enters only in the second order. Even here, by following the detailed series expansions given by Danby[7] or Moulton[8], one can generally reduce the error

105

in the C_ks and hence the orbital elements to values consistent with the errors of observation. From a rather protracted argument Danby[7] gives the following expressions for the two linearly independent C_is:

$$\left.\begin{array}{l} C_1 = \dfrac{(t_3 - t_2)}{(t_3 - t_1)} \left\{ 1 + \dfrac{k^2}{6r_2^2} \left[(t_3 - t_1)^2 - (t_3 - t_2)^2 \right] \right\} \\[3em] C_3 = \dfrac{(t_2 - t_1)}{(t_3 - t_1)} \left\{ 1 + \dfrac{k^2}{6r_2^2} \left[(t_3 - t_1)^2 - (t_2 - t_1)^2 \right] \right\} \end{array}\right\} \qquad (7.2.21)$$

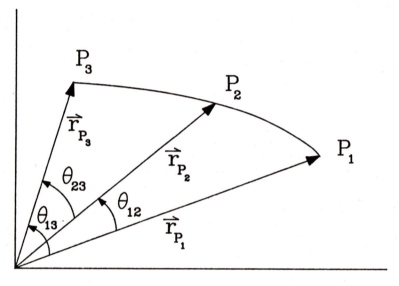

Figure 7.3 shows a section of the orbit of an object revolving about the sun. The object is observed at three points P_i in its orbit and the method of Gauss determines the three heliocentric radius vectors $\vec{r}_{pi}$. The area A_{ij} is the area of the triangle made from the heliocentric radius vectors $\vec{r}_{pi}$.

The improvements in the estimations of the C_is involve information about the orbit in the form of the factors of $(k^2/6r_2^2)$, as they must, because they involve the corrections required to go from the orbital sectors bounded by the

106

heliocentric radius vectors to the triangles that they form. Danby[7] gives an improved method due to Gibbs which provides a somewhat more accurate form of the approximation, but the concept is the same.

7.3 Degeneracy and Indeterminacy of The Orbital Elements

Before leaving the discussion of orbital elements, I would like to emphasize further that the equations of celestial mechanics are nonlinear. Of the many problems this exacerbates, one is the determination of the orbital elements for the object. Occasionally two of the orbital elements become redundant or indeterminate depending from which end of the two body problem one is starting. For a circular orbit clearly there is no perihelion or point of closest approach. Therefore, there can be no time of perihelion passage. Similarly if the orbital inclination is zero, the orbital plane is co-planar with the plane that defines the coordinate frame and there will be no line of nodes. In this instance the longitude of the ascending node is undefined. One may define the problem out of existence by simply taking the passage of the first point of Aries or the vernal equinox as the reference point for measuring time and the true anomaly. If the inclination is zero and the orbit elliptical, one could simply measure the argument of perihelion from the vernal equinox and have a perfectly well defined orbit, and no trouble would be encountered in locating the object in the sky.

However, in the event that one is determining the orbital elements from observation, there is no advanced information regarding the pathology of the solution. If the inclination is small, the error in the longitude of the ascending node Ω will be large. Similarly, should the eccentricity prove to be very small, the error in the argument of perihelion will be large, so that the time of perihelion passage is poorly known. These errors propagate in a highly nonlinear way and one must be ever mindful of them. The problems caused by a low value of the inclination are not fundamental but result from an unfortunate choice of the coordinate system. They can be eliminated by choosing a different coordinate frame in which to do the calculations. However, the problems are real and will return upon subsequent transformation to the original coordinate frame. The problems introduced by circular orbits are more fundamental as they result from a degeneracy of the orbit itself, and that cannot be transformed away. One can take some comfort from the fact that an uncertain location of the point of perihelion does not mean that the location of the object in its orbit will be uncertain since that error is usually compensated by an opposite error in the time of perihelion passage. The errors in the orbital elements will not be linearly independent so that the net result in locating the object in its orbit will not necessarily be serious. It is better under these conditions to measure the time in the orbit from some well determined location such as the vernal equinox.

In this chapter we have seen how to determine the orbital elements of an object from observational information regarding its motion. This constitutes the second part of the classical celestial mechanics problem of describing the motion of one object about another. In practice, the calculation of precise orbital elements involves many additional practical details concerned with both observations and the theory, but the overall approach is roughly that described here. There are a number of alternative approaches to finding the orbital elements. Indeed, is said that Gauss devised some thirteen different schemes for his doctoral thesis. However, the information content of three sets of angular measurements or the equivalent is always required and the details concern only the devoted practitioner. In the previous chapter we used the elements to predict the motion of the object on the sky. Thus, the two pieces can be put together to predict the motion of an object on the basis of observations of its motion. This is certainly the classical task of any science - that is, to predict the future behavior of the physical world from knowledge of its current behavior. This was a great triumph for Newtonian mechanics in the 17th and 18th centuries and indeed for science itself. The mathematicians and philosophers who came after Newton developed this elegant determinism to deal with much more formidable problems than the two body problem. For the remainder of the book we shall look at some of their successes and some of the remaining problems.

Chapter 7: Exercises

1. Find the altitude, azimuth, Right Ascension, and Declination of the planet Venus as seen from Columbus Ohio at 9:00 PM EST February 10, 1988.

    Given the orbital elements:

    $a = 0.7233316$ AU
    $e = 0.006818 - 0.00005T$
    $\ell = 81°\ 34'\ 19"$ (on Jan 0.5, 1950)
    $i = 3°\ 23'\ 37\overset{..}{}1 + 4\overset{..}{}5 \times T$
    $P = 0.6151856$ yr. $= 224.701$ days
    $\Omega = 75°\ 47'\ 01" + 3260" \times T$
    $\tilde{\omega} = 130°\ 09'\ 08" + 5065" \times T$
    $T = (67 + Date/365.25)/100$

2. With what geocentric velocity must an artificial satellite be launched horizontally from the earth in order that its apogee distance from the earth's center be 60 earth radii (approximately the moon's distance)? What will be the orbital eccentricity and the orbital period? Ignore air resistance and the gravitational effects of other bodies in the solar system.

3. You plan a trip to Venus. Assume that the orbits of the earth and Venus are circular and co-planar. You will launch your ship from the earth in a direction directly opposite to the earth's orbital motion so that spacecraft has velocity V with respect to the sun when it is "clear of the earth". Note that $V < V_E$ (the earth's orbital speed) and the ship is at its aphelion point at launch. We desire the perihelion point to be at the orbit of Venus ($a=0.723$AU). What are the semi-major axis (a_r) and the eccentricity (e_r) of the spacecraft's transfer orbit in terms of V? What is the orbital period of the spacecraft? What is the travel time to Venus and where should Venus be in the sky at the time of launch in order to ensure its presence when you arrive?

4. If the heliocentric *cartesian* coordinates (i.e., the origin is at the sun, the x-axis points toward the vernal equinox, and the z-axis is normal to the ecliptic plane) of a certain comet on November 26.74, 1910 were:

$$\vec{r} = [(+2.795526), (+1.399919), (0)]$$

and,

$$\omega = 267° \ 16' \ 36\overset{.}{''}6$$
$$\Omega = 260° \ 40' \ 11\overset{.}{''}8$$
$$i = \ 18° \ 29' \ 41\overset{.}{''}1 \quad ,$$

find the heliocentric coordinates on January 0.5, 1986. Find the altitude and azimuth as seen from Columbus Ohio at 7:00AM EST. Ignore the difference between ET and UT.

5. Given the heliocentric equatorial position and velocity vectors of an object in orbit about the sun to be

$$\left. \begin{array}{l} \vec{r}_0 = 2\hat{i} + 2\hat{j} + \hat{k} \\[2mm] \dot{\vec{r}}_0 = \hat{i}/3 + \hat{j}/3 + \hat{k} \end{array} \right\} \quad ,$$

where the units of time and distance are years and astronomical units. Find the position of the object two years later. List specifically all assumptions you make and describe clearly the approach you took to the problem.

110

8

The Dynamics of More Than Two Bodies

In chapter 3 we established the general principles of Newtonian mechanics and the mathematical formalism for the determination of the equations of motion for the objects that make up an arbitrary mechanical system. We used those principles in chapter 5 to describe the motion of two bodies under their mutual gravitational attraction. As we shall see, problems dealing with more than two bodies become extremely complicated and do not, in general, yield closed form solutions. The dynamical behavior of large systems of stars that seem to populate the central regions of galaxies is currently a problem of intense study three and a half centuries after Newton identified the principles that guide their motion. Before we even attempt to discuss systems consisting of a large number of objects, we shall discuss systems of three objects.

8.1 The Restricted Three Body Problem

Certainly the next logical step after the solution of the two body problem is the addition of a third body. Yet even here we find that the general problem is unsolved. Nature seems to deal with the problem in a simple manner for there are many stellar systems consisting of three or more stars bound by their mutual gravitational attraction. However, in all of these systems, the objects seem to degenerate to a hierarchical succession of two body problems. For example, should the system contain three stars, two will be tightly bound orbiting as one would expect from the two body solution and the third will be found at a distance corresponding to many times the separation of the close orbiting pair. Four gravitationally bound stars always appear as a binary of binaries and so forth. It is generally believed that there are no stable orbits involving three comparable masses with comparable separations.

The source of the difficulty in dealing with as few as three objects can be found in the notion of the integrals of the motion discussed in chapter 5 (Sec 5.3). In any mechanics problem one always has the Hamiltonian, or the total energy, and the total angular momentum as integrals of the motion. These are quantities which will be constant throughout the motion of the members of the system wherever they may go. Since the angular momentum is a vector quantity, it has three linearly independent components, each of which serves as a constant of the motion. Thus the conservation of energy and angular momentum provide four constants that restrict the motion of the system. Taken together with the six constants that specify the uniform motion of the center of mass, there remain only two constants to completely determine the motion of a two body system. It is the quadratic nature of the force law that requires that solutions for the orbits will also be quadratic and thus if the orbits are bound they will be closed. This is not the case for other force laws, as is evidenced by the precession of the perihelion of Mercury's orbit resulting from the presence of masses other than the sun in the solar system. Mercury's orbit isn't quite elliptical and never exactly closes in space. Closure requires that the object return to the same physical location with the same velocity. Thus the last constant serves only to locate the particle in its orbit.

Since the general problem of three bodies will be described by a second order vector differential equation for each of the particles, there will be 18 constants of motion. The conservation of angular momentum and energy together with the uniform motion of the center of mass will provide 10 constants leaving eight to be determined. Since the general potential affecting any one of the objects will not be that of a single point mass we should not expect the orbits of the objects to close and we are left with eight arbitrary constants required to specify the problem. Thus the motion is in no way uniquely determined by the conservation laws of physics as was the case for the two body problem. To be sure the initial position and velocities of the components would provide the 18 constants required for the unique solution of the motion since the laws of Newtonian mechanics are deterministic. But these initial values are not integrals of the motion. The parameters they specify are not constant during the motion of the members of the system. Thus while they provide a basis for calculating the motion of a specific system, they do not allow for a general solution. Since the general solution of the three body problem appears beyond our grasp, let us consider a simpler problem intermediate between the two body problem and the general three body problem.

The question of what is the most complicated problem in celestial mechanics that allows for a general solution has occupied some of the best analytical minds of the past three centuries and continues to be of interest today. Consider two bodies of comparable but dissimilar masses in circular orbit about one another. Now introduce a third object of negligible mass. Here "negligible mass" means that it is affected by the presence of the other two objects, but does not exert sufficient force on either of the two so as to disturb their circular motion. It is then a reasonable question to inquire into the motion of this third object. Such a question is not entirely academic as this is an excellent approximation to the motion of a spacecraft in the earth-moon system. It is also a fair approximation to the motion of some asteroids influenced by the gravitational fields of the sun and Jupiter. This problem is called the circular restricted three body problem and its solution contains some surprising results.

a. Jacobi's Integral of the Motion

We analyzed the two body problem in physical units, but we are free to choose any system of units we please. So let us measure mass in units such that the total mass of the system is unity. Then

$$m_1 + m_2 = M \equiv 1 \quad . \qquad (8.1.1)$$

We could then quite arbitrarily require the less massive of the orbiting pair to have a mass μ so that

$$\left. \begin{array}{l} m_1 \equiv \mu \\[2ex] m_2 \equiv 1 - \mu \\[2ex] \mu \leq \tfrac{1}{2} \end{array} \right\} \qquad (8.1.2)$$

Remember that the third object in the system has essentially zero mass in that it doesn't contribute to the total mass of the system at any level that could be considered significant. Indeed, it behaves as a 'test particle' as described in chapter 5. Now we are free to choose the units by which we measure time so instead of using seconds, let us measure time in units of the orbital period of the two significant objects about one another. For the earth and the sun this would be years multiplied by 2π. Such a choice requires that the attractive force between the objects be such that

$$\omega \equiv k[(1-\mu)+\mu]/d^3]^{\tfrac{1}{2}} \equiv 1 \quad . \qquad (8.1.3)$$

113

Now for the description of the motion of the third object, let us choose a cartesian coordinate system with an origin at the center of mass and rotating with the uniform circular motion of the two non-negligible masses. Thus the least massive object will be located at x_1 and the more massive one at x_2. The third object will have coordinates [x,y,z] so that its radial distance from the two objects can be represented by

$$\left.\begin{array}{l} r_1^2 = (x-x_1)^2 + y^2 + z^2 \\[1em] r_2^2 = (x-x_2)^2 + y^2 + z^2 \end{array}\right\} \qquad . \tag{8.1.4}$$

In an inertial coordinate system the total energy would simply be

$$\tfrac{1}{2}\Sigma_i m_i v_i^2 - Gm_1 m_2/d = E \quad . \tag{8.1.5}$$

However, if the coordinate system is rotating, the kinetic energy will be reduced by the rotational motion and, to conserve energy, we will have to increase the potential energy by a corresponding amount. Since the orbits of m_1 and m_2 are circular, their contribution to the kinetic and potential energies of the system will separately remain constant. Thus energy conservation can be reduced to the energy of the small mass body constant. If we let the object have a mass ϵ, then the total energy of the small body is

$$\tfrac{1}{2}\epsilon v^2 - \tfrac{1}{2}\epsilon\omega^2(x^2+y^2) - \epsilon(1-\mu)/r_1 - \epsilon\mu/r_2 = const. = E(\epsilon). \tag{8.1.6}$$

Here v is the velocity measured in the rotating coordinate frame and the quantity $\tfrac{1}{2}\epsilon\omega^2(x^2+y^2)$ is just the contribution from the rotational motion of the coordinate frame itself. Dividing out the negligible mass of the third body and taking $\omega=1$, we can write

$$v^2 = \omega^2(x^2+y^2) + 2(1-\mu)/r_1 + 2\mu/r_2 - C \quad , \tag{8.1.7}$$

where C is some constant of the motion. This is known as Jacobi's integral and is nothing more than the energy integral for the third body. Now it is clear why the orbits of the other two bodies were assumed to be circular. Still the equations of motion for the third object require six constants of motion for complete specification of the motion of the third body. Thus we need five more. The total angular momentum of the system is conserved, but it is entirely tied up in the motion of the two objects and thus is of little help here. The remaining five constants are simply not known, so that it is remarkable that we may say anything about the motion of the third object.

114

b. Zero Velocity Surfaces

Now $v^2 \geq 0$ by definition so that Jacobi's integral places limits on where the third object may go depending on the value of C. Let us consider surfaces where $v = 0$. These are surfaces that provide bounds for the third object's motion, for the particle cannot cross them. For it to do so the square of the particle's velocity would have to change sign. Remembering that we can take $\omega = 1$, we can write the expression for the zero velocity surfaces as

$$(x^2+y^2) + 2(1-\mu)/r_1 + 2\mu/r_2 = C \quad . \tag{8.1.8}$$

Clearly the value of C must always be positive. Therefore consider the case where $C \gg 0$. Then either $C \approx (x^2+y^2)$ or one of the radial distances $[r_1, r_2]$ must be small. Thus either the third body is very close to one of the objects in a tight orbit about it or it is very distant and moves as if the pair were a point source. This is the solution most commonly found in nature. The zero-velocity surfaces would then consist of a cylinder normal to the x-y plane at some distance rather greater than the separation distance d and two smaller 'egg-shaped' surfaces close to each of the objects. These surfaces confine the motion to outside the cylinder or within the oval surfaces. As the value of C is decreased, the outer cylinder decreases in radius and the inner ovoids become bigger. As the value of C continues to decrease the two inner ovoids will touch at a point along the line joining two circularly revolving objects. Let this value be called C_2 and the physical point in space labeled L_1. A particle confined within the ovoids will then be able to move from one to another as this "double point" no longer divides regions of space where v^2 has opposite sign. As the ovoids continue to grow with decreasing C they join at L_1 forming a hour glass shaped structure that grows to meet the shrinking cylinder. Eventually, as C takes on smaller and smaller values, the two regions will meet first along the line joining the centers and behind the less massive of the two principle masses. Let this value of C be called C_3 and the corresponding spatial location be known as L_2. The point that occurs behind the more massive of the two objects is known as L_3 and occurs when C decreases to C_4. A further decrease in the value of C causes the surfaces to separate into two comma shaped regions in the x-y plane which asymptotically approach two points when C becomes C_5. These two points can be distinguished in that one leads the more massive object in its orbit while the other trails behind. They are called L_4 and L_5 respectively. The L_is are collectively called the Lagrange points and have special significance. Figure 8.1 shows cross sections of these surfaces in the x-y and x-z planes.

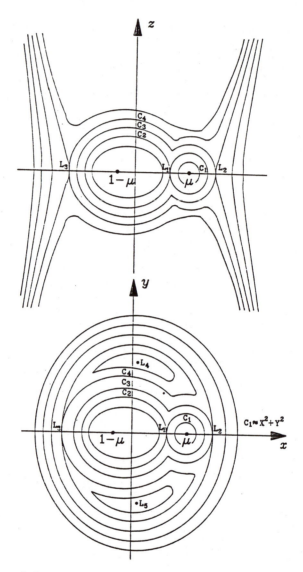

Figure 8.1 shows the zero velocity surfaces for sections through the rotating coordinate system. The upper drawing shows the cross section through the x-z plane while the lower drawing shows the cross section of the x-y plane. The various values of C, as well as the location of the Lagrangian points of equilibrium, are indicated.

c. The Lagrange Points and Equilibrium.

In chapter 5 [equation (5.3.4)] we defined a "rotational potential" to account for the centrifugal forces generated by the conservation of angular momentum. In a similar manner, we can define a new potential to take account of the rotation of the coordinate frame by including the energy resulting from the motion of the coordinate frame itself. Let this potential be

$$\emptyset = - \tfrac{1}{2}\omega^2 (x^2 + y^2) - (1-\mu)/r_1 - \mu/r_2 \quad , \qquad (8.1.9)$$

so that the total energy of the third body is

$$E(\epsilon) = \tfrac{1}{2}\epsilon v^2 + \epsilon\emptyset \quad . \qquad (8.1.10)$$

The forces acting on the third body will just be

$$\vec{F}(\epsilon) = - \nabla\emptyset = - \nabla(\emptyset - \tfrac{1}{2}C) = - \nabla(\tfrac{1}{2}v^2) \quad . \qquad (8.1.11)$$

Since the function v^2 vanishes at the points where two zero velocity surfaces meet and $v^2 \geq 0$, its gradient must also vanish on those surfaces. Thus the points of tangency represent places of equilibrium where all forces on the third body vanish. It remains to be established if those points represent stable equilibrium. Therefore the Lagrangian points may be found from

$$\tfrac{1}{2}\nabla^2 (v^2) = 0 = \hat{i}v(\partial v/\partial x) + \hat{j}v(\partial v/\partial y) + \hat{k}v(\partial v/\partial z) \quad . \qquad (8.1.12)$$

Since each component of the vector must be zero separately, the equations of condition for the Lagrangian points are

$$
\left.
\begin{aligned}
v\frac{\partial v}{\partial x} &= x - \frac{(1-\mu)(x-x_1)}{r_1^3} - \frac{\mu(x-x_2)}{r_2^3} = 0 \\[2ex]
v\frac{\partial v}{\partial y} &= y - \frac{(1-\mu)y}{r_1^3} - \frac{\mu y}{r_2^3} = 0 \\[2ex]
v\frac{\partial v}{\partial z} &= z\left[\frac{(1-\mu)}{r_1^3} + \frac{\mu}{r_2^3}\right] = 0
\end{aligned}
\right\} \qquad (8.1.13)
$$

Neither r_1 nor r_2 is zero and $\mu \neq 1$ so that the z-component of the gradient requires that $z = 0$ and that all the Lagrangian points lie in the x-y plane. If $y \neq 0$, then the y-component of the gradient requires that

$$1 - (1-\mu)/r_1^2 - \mu/r_2^2 = 0 \quad . \tag{8.1.14}$$

This has a solution for

$$r_1 = r_2 = 1 \quad . \tag{8.1.15}$$

Since in the units we are using the separation of the two orbiting masses is unity, these points must lie at the vertices of equilateral triangles in the x-y plane having the line joining the two orbiting masses as a base. These are the points L_4 and L_5. Thus the Lagrangian points L_4 and L_5 lie in the orbital plane, leading and following the orbiting bodies by 60° at a distance equal to the separation of those two bodies. If we satisfy the conditions on the gradient by requiring both y and z to be zero, then the x-component of the gradient requires

$$x - \frac{(1-\mu)(x-x_1)}{|x-x_1|^3} - \frac{\mu(x-x_2)}{|x-x_2|^3} = 0 \quad , \tag{8.1.16}$$

and the remaining Lagrangian points will lie along the x-axis at the roots of the polynomic equation (8.1.16). In general, all solutions must be found numerically. However, Moulton[8] (p.290) gives series solutions for the location of the Lagrangian points in terms of μ.

In order to test the nature of the stability of the Lagrangian points one need calculate

$$\frac{\partial^2 \varnothing}{\partial x_i^2} < 0 \quad . \tag{8.1.17}$$

If this condition is satisfied for all of the coordinates, then each of the points represents a point of stable equilibrium. That is, if a particle is slightly displaced from the point, the particle will return to it. This is the case for L_4 and L_5. However, L_1, L_2, and L_3 are all unstable and an object displaced from any one of them will continue to move away from them. Since the condition given in equation (8.1.17) is essentially the derivative of the forces acting on the particle, stable equilibrium requires that a small displacement generate a small negative force pushing the object back where it came from. A small positive force would continue to accelerate the particle away from its earlier location. The relative stability of the Langrangian points can be seen from Figure 8.1. For L_1, L_2, and L_3 the touching of the zero velocity surfaces joins two regions where the motion of the particle was previously confined. Thus particles can freely roam from one region to the other. A particle at one of these points could then move either way and would not be stable. However, Lagrangian points L_4, and L_5 represent the 'center' of a forbidden region where $v^2 < 0$ so that the kinetic energy would have to be increased in order to move away from them. As the value of C is decreased so that the forbidden region shrinks to a point, that point can be occupied, but only by a particle with zero velocity. A small displacement would not provide the kinetic energy required for the particle to return to the point and the point would be stable.

The Lagrangian points are important in astronomy as they mark places where particles can either be trapped (L_4 and L_5) or will pass through with a minimum expenditure of energy. In the solar system there are two sets of asteroids known as the Trojan asteroids that lead and follow Jupiter about in its orbit oscillating about L_4 and L_5. In the theory of binary star evolution, the more massive component will expand as it ages until material meets one of the Lagrangian points. If that point is L_1, the matter will stream across the gap between the two stars and eventually be accreted onto the other member of the system. Should either L_2 or L_3 be encountered, the matter will pass through and is likely to be lost to the system entirely.

Much more could be said (eg. Moulton[8]) about the restricted three body problem as books have been written on the subject and some people have devoted their lives to its study. However, its most important aspects are bound up in the study of Jacobi's Integral and it is remarkable that so much can be said about the motion of the third body from knowledge of one integral of the motion.

8.2 The N-Body Problem

After encountering the difficulties posed by the three body problem it must seem foolhardy to even consider larger systems. However, the universe is full of systems of many objects that are largely bound by their mutual gravity and we would like to understand as much about their dynamics as possible. Let us begin by determining the equations of motion for such a system. We can do this as we did for central forces and the two body problem by calculating the Lagrangian. Thus,

$$\mathcal{L} = \sum_i \tfrac{1}{2} m_i (\dot{\vec{r}}_i \cdot \dot{\vec{r}}_i) + \tfrac{1}{2} \sum_{i \neq j} \sum_j (Gm_i m_j / d_{ij}) \qquad . \qquad (8.2.1)$$

The equations of motion are therefore

$$m_i \ddot{\vec{r}}_i = -\tfrac{1}{2} Gm_i [\sum_{j \neq i} m_j (\vec{r}_i - \vec{r}_j) / d_{ij}^3] \qquad . \qquad (8.2.2)$$

119

Summing these equations over all the particles we get

$$\sum_i m_i \ddot{\vec{r}}_i = -\tfrac{1}{2} G \sum_i \sum_{j \neq i} m_i m_j (\vec{r}_i - \vec{r}_j)/d_{ij}^3 \quad . \tag{8.2.3}$$

Now since

$$\vec{r}_i - \vec{r}_j = -(\vec{r}_j - \vec{r}_i) \quad , \tag{8.2.4}$$

we may pair the terms in the double sum on the right hand side of equation (8.2.3) so that they individually cancel to zero leaving

$$\sum_i m_i \ddot{\vec{r}}_i = 0 \quad . \tag{8.2.5}$$

This equation can be directly integrated twice with respect to time to get

$$\sum_i m_i \vec{r}_i = \vec{A}t + \vec{B} \quad . \tag{8.2.6}$$

The left hand side of this equation is the definition of the center of mass and the vectors on the right hand side have six linearly independent components. Thus, even for a dynamical system of N particles, the center of mass will move with a uniform velocity. However, N second order vector equations will require 6N constants of integration in order to uniquely specify the motion of the particles and finding six seems of little help.

Taking the cross product of a position vector with the equations of motion we can write

$$\vec{r}_i \times m_i \ddot{\vec{r}}_i = -\tfrac{1}{2} G m_i \sum_{i \neq j} m_j [\vec{r}_i \times \vec{r}_i - \vec{r}_i \times \vec{r}_j]/d_{ij}^3$$

$$= \tfrac{1}{2} G \sum_{i \neq j} [\vec{r}_i \times \vec{r}_j]/d_{ij}^3 \quad . \tag{8.2.7}$$

Since

$$\vec{r}_i \times \vec{r}_j = -(\vec{r}_j \times \vec{r}_i) \quad , \tag{8.2.8}$$

we can again sum equation (8.2.7) over all the particles and pair the terms under the double sum of the right hand side so that they vanish to zero. Thus we may write

120

$$\sum_i \vec{r}_i \times m_i \ddot{\vec{r}}_i = \sum_i \vec{r}_i \times \dot{\vec{p}}_i = \frac{d}{dt} \sum_i (\vec{r}_i \times \vec{p}_i) = \frac{d}{dt} \sum_i \vec{L}_i = \frac{d\vec{L}}{dt} = 0 \ , \qquad (8.2.9)$$

and find that the total angular momentum of all the particles will be constant. Thus we add three more constants of the motion to our total. We also establish that there will be a fundamental plane of the system that is perpendicular to the total angular momentum vector. Similarly we can invoke the conservation of the total energy to get a last constant of the motion as

$$\tfrac{1}{2}\sum_i m_i \dot{r}_i^2 - \tfrac{1}{2}\sum_j \sum_{i \neq j} Gm_i m_j / d_{ij} = E = \text{const.} \qquad (8.2.10)$$

Thus, as was the case with the three body problem, we have 10 integrals of the motion, far short of the 6N needed to complete the solution. In addition, all of these constants of the motion are global. That is, they refer to properties of the total system and tell us little about the motion of individual particles. However, there is one more global condition that is of considerable help in understanding the history of the system.

a. The Virial Theorem

The virial theorem, as it is commonly called in the literature, takes on many forms. However, all of them have in common a relationship whose origins are in the equations of motion for the system. We will generate only the simplest of these relationships, namely that appropriate for particles moving under the influence of the gravitational force. A derivation for an arbitrary central force law is given by Collins[9]. The general equations of motion for such a system of particles are given in equation (8.2.2). Now take the scalar product of those equations with a position vector to each object in the system and sum over all the particles so that

$$\sum_i m_i \ddot{\vec{r}}_i \cdot \vec{r}_i = \sum_i m_i \frac{d}{dt}(\dot{\vec{r}}_i \cdot \vec{r}_i) - \sum_i m_i \dot{\vec{r}}_i \cdot \dot{\vec{r}}_i$$

$$= G \sum_i m_i \sum_{j \neq i} m_j (\vec{r}_i - \vec{r}_j) \cdot \vec{r}_i / d_{ij}^3 \qquad (8.2.11)$$

We can rewrite the central part of equation (8.2.11) so that

$$\tfrac{1}{2}\frac{d^2}{dt^2}\left[\sum_i m_i r_i^2\right] - 2T =$$

$$G\sum_i\left\{\sum_{j>i} m_i m_j\left[(\vec{r}_i-\vec{r}_j)\cdot\vec{r}_i+(\vec{r}_j-\vec{r}_i)\cdot\vec{r}_j\right]\Big/d_{ij}^3\right\} \quad (8.2.12)$$

where T is the total kinetic energy of the system. We have also rewritten the left hand side of equation (8.2.11) to explicitly show the pairing of terms for the force of ith particle on the jth particle with the force of the jth particle on the ith particle. The first term in square brackets is a "moment of inertia"-like term only instead of it being a moment of inertia about an axis it is the moment of inertia about the origin of the coordinate system. Let us call this quantity I so that equation (8.2.12) becomes

$$\tfrac{1}{2}\frac{d^2 I}{dt^2} - 2T = G\sum_i\sum_{j>i} m_i m_j\left[(\vec{r}_i-\vec{r}_j)\cdot(\vec{r}_i-\vec{r}_j)\right]/d_{ij}^3$$

$$= G\sum_i\sum_{j>i}\frac{m_i m_j}{d_{ij}} \quad . \quad (8.2.13)$$

The term on the far right is the negative of the potential energy of the system so that

$$\tfrac{1}{2}\frac{d^2 I}{dt^2} = 2T - V \quad . \quad (8.2.14)$$

Some call this the virial theorem, but it is more correctly known as Lagrange's identity even though Lagrange only proved it for the case of three bodies. Karl Jacobi generalized it to a system with N-bodies and it is clearly an identity. That is, it is very like a conservation law as it must always be true for any dynamical system. Now if one integrates Lagrange's identity over time, one can write for stable or bound systems that

$$\text{Lim}_{t \to \infty} \left[\frac{1}{t} \int \frac{1}{2} \frac{d^2 I}{dt^2} \, dt \right] = 0 = 2\langle T \rangle - \langle V \rangle \quad . \qquad (8.2.15)$$

This result also holds if the system is periodic and the integral is taken over the period of the system, since the system must return to an earlier state so that the moment of inertia and all its derivatives are the same at the limits of the integral. Equation (8.2.15) is known as the time-averaged form of the virial theorem (or generally just the virial theorem) and provides an additional constraint on the behavior of the system.

b.   The Ergodic Theorem

The ergodic theorem is in the category of concepts that are so basic that they are never taught, but are assumed to be known. For that reason alone, it is worth discussing. Ergodic theory constitutes a major branch of mathematics and its physical application has occupied some of the best minds of the twentieth century. The theorem from which this branch of mathematics takes its name basically says that the average of some property of a system over all *allowed* points in phase space is identical to the average of that same quantity over the entire lifetime of the system. To explain this and its implications, we must first say what is meant by phase space.

Consider a 6-dimensional space where the coordinates are defined as the location and momentum of a particle. The coordinates of the particle in such a space specify its position and momentum, which requires the six components of its location in phase space. These six components constitute the six constants required for the solution of the Newtonian equations of motion. Thus locating a particle in phase space fixes its entire history - past and future and thus determines the path that the particle will take through phase space as it moves. However, not all points in phase space are allowed to the particle, for its total energy cannot change and there are points in phase space that correspond to different total energies for the particle. Thus, the path of the particle in phase space will be limited to a "space" of one lower dimension - namely one where the total energy is constant. Quantities that limit the phase space available to the motion of a particle are said to be *isolating integrals of the motion* and certainly the total energy is one of them. If we are dealing with the motion of a single particle in an arbitrary conservative force field, its angular momentum will also be an isolating integral.

123

Thus, the ergodic theorem says that a particle will reach every point in the phase space allowed to it during its lifetime so that the path of the particle will completely cover the space. Therefore averages of quantities taken over the space are equivalent to averages taken over the lifetime. This seemingly esoteric theorem is of fundamental importance to physics. In thermodynamics we make predictions about time averages of systems but can observe only phase space averages. Thus to relate the two, it is necessary to invoke the "Ergodic hypothesis" - namely, that the ergodic theorem applies to thermodynamic systems. The best justification for this hypothesis is that thermodynamics works!

Unfortunately, the ergodic theorem has never been proved in its full generality, but sufficiently general versions of it have been proved so that we may use it in science. This allows us to replace the time averages that appear in equation (8.2.15) with averages over phase space. This is fortunate as the average astronomer doesn't live long enough to carry out the time averages required to use the virial theorem. The difficulty in applying the virial theorem is in deciding exactly in what subspace is the system ergodic; that is, in deciding how many isolating integrals of the motion there are and what are they. Without that information, we cannot determine how to carry out the averages over the appropriate phase space.

What sorts of things might we want to average? Clearly for the virial theorem we would like to know the average of the kinetic and potential energies for if they do not satisfy equation (8.2.15), the system is not stable and will eventually disperse. Conversely, if the system is adjudged to be a stable system, the average of one of these quantities, together with the virial theorem, will provide the other. This is often used to determine the mass of stable systems.

c. Liouville's Theorem

We conclude our discussion of the N-body problem with a brief discussion of a theorem that deals with the history of an entire system of particles. To do this, we need to generalize our notion of phase space. Consider a space of not just six dimensions, but 6N dimensions where N is the number of particles in the system. Each of the dimensions represents either the position or momentum of one of the particles. As there was need of six dimensions for a system consisting of one particle, the 6N dimensions will suffice to specify the initial conditions for every particle in the system. Thus, the system represents only a point in this huge space, and the space itself is the space of all possible systems of N particles.

Such a space is usually distinguished from phase space by calling it configuration space. The temporal history of such a system will be but a single line in configuration space. Liouville's theorem states that the density of points in configuration space is constant. This, in turn, can be used to demonstrate the determinism and uniqueness of Newtonian mechanics. If the configuration density is constant, it is impossible for two different system paths to cross, for to do so, one path would have to cross a volume element surrounding a point on the other path thereby changing the density. If no two paths can cross, then it is impossible for any two ensembles ever to have exactly the same values of position and momentum for all of their particles. Equivalently, the initial conditions of an ensemble of particles uniquely specify its path in configuration space. This is not offered as a rigorous proof, but only as a plausibility argument. More rigorous proofs can be found in any good book on classical mechanics.

The ergodic theorem applies here as well, for if any two systems ever cross in configuration space, they must in reality be the same system seen at different times in its dynamical history. Clearly the paths of systems with different total energies can never cross in accord with Liouville's theorem, but will cover the subspace allowed to them in accordance with the ergodic theorem.

These three theorems are powerful products of the great development of classical mechanics of the nineteenth century. They give us additional and rigorous constraints that apply to systems with any number of particles and they lie at the very foundations of modern physics. They are basically statements of conservation laws and the determinism of Newtonian physics.

8.3 Chaotic Dynamics in Celestial Mechanics

Theoretical physics has had a difficult time, in general, describing phenomena that exhibit some degree of order, but not complete order. Total disorder can be dealt with and thermodynamics is an example of highly developed theoretical structure that deals with gases whose constituents show totally random behavior. Classical mechanics describes well ordered systems with great success. However, intermediate cases are not well understood. This weakness in theoretical physics can be found throughout the discipline from the theories of radiative and convective transfer of energy, to "cooperative phenomena" in stellar dynamics. We have seen from our study of the N-body problem that non-periodic solutions and ergodic paths in phase space can result. The solar system is an N-body system, yet it clearly displays a high degree of order. Might not we expect some aspects of it to behave otherwise?

The space program of the 1960s and 1970s brought us detailed photographs of various objects in the solar system whose dynamical behavior proved to be far more complicated than was previously imagined. The rings of Saturn proved to be more numerous and structured than anyone believed possible. One of the Saturnian satellites (Hyperion) appears to tumble in an unpredictable manner. The rings of Uranus have a structure that most astronomers would have thought was unstable. This list is far from exhaustive, but begins to illustrate that there are many problems of celestial mechanics that remain to be solved. One of the most productive approaches to some of these problems has been through the developing mathematics of Chaos. In the area of dynamics, chaotic phenomena are those that, while being restricted in phase space, do not exhibit any discernable periodicity. Wisdom[10] has written a superb review article on the examples of chaos in the dynamics of the solar system and we will review some of his observations.

In the nineteenth century Poincare showed that integrals of the motion usually do not survive orbital perturbations. Thus, closed form integration of perturbed orbits will not, in general, be possible. However, a more recently proved theorem known as the KAM theorem shows that for small perturbations, orbital motion will remain quasi-periodic. Thus the simple loss of the integrals of the motion does not imply that the dynamical motion of the object will become unrestrained in phase-space and be ergodic. This somewhat surprising result implies that we might expect to find orbits that are largely unpredictable but remain confined to parts of phase space. Wisdom points out that the phase space accessible to a system with a given Hamiltonian may depend critically on the initial conditions. For some sets of initial conditions, the motion of the system will be quasi-periodic, and the system will be confined to a relatively small volume of phase space. For modest changes in the initial conditions, the motion of the system becomes chaotic and completely unpredictable. It is a characteristic of such systems that the transition from one region to another is quite sharp. A similar situation is seen in thermodynamics when a system undergoes a phase transition. Here the mathematics of Chaos has been relatively successful in describing such transitions.

A simple example of such a dynamical system can be found in the restricted three body problem. From Figure 8.1 it is clear that an object orbiting close to one or the other of the two main bodies will experience nearly elliptical motion that is certainly quasi-periodic. However, for values of C of the order of C_3 the motion is barely confined and numerical experiments show that the orbits wander over a large range of the allowable phase space in a non-periodic manner. Thus with chaotic behavior being present in such a relatively simple

126

system, we should not be surprised to find it in the solar system. While analysis of such systems is still in its infancy we know enough about the mathematics of Chaos to be confident that it will lead to a more complete understanding of non-linear dynamical systems. We are once again reminded that the future of theoretical physics can be seen "through a glass darkly" in the developing mathematics of the present.

Chapter 8: Exercises

1. Show that the Lagrangian points L4 and L5 are points of stable equilibrium while the Lagrangian points L1-L3 are not.

2. Derive the virial theorem for an attractive potential that varies as $1/r^2$.

3. Show that the virial theorem has its normal form even if there are velocity dependent forces present.

4. How many isolating integrals of the motion are there for the case of just two orbiting bodies? What does this mean for the application of the ergodic theorem to the virial theorem?

9

Perturbation Theory and Celestial Mechanics

In this last chapter we shall sketch some aspects of perturbation theory and describe a few of its applications to celestial mechanics. Perturbation theory is a very broad subject with applications in many areas of the physical sciences. Indeed, it is almost more a philosophy than a theory. The basic principle is to find a solution to a problem that is similar to the one of interest and then to cast the solution to the target problem in terms of parameters related to the known solution. Usually these parameters are similar to those of the problem with the known solution and differ from them by a small amount. The small amount is known as a perturbation and hence the name perturbation theory.

This prescription is so general as to make a general discussion almost impossible. The word "perturbation" implies a small change. Thus, one usually makes a "small change" in some parameter of a known problem and allows it to propagate through to the answer. One makes use of all the mathematical properties of the problem to obtain equations that are solvable usually as a result of the relative smallness of the perturbation. For example, consider a situation in which the fundamental equations governing the problem of interest are linear. The linearity of the equations guarantees that any linear combination of solutions is also a solution. Thus, one finds an analytic solution close to the problem of interest and removes it from the defining equations. One now has a set of equations where the solution is composed of small quantities and their solution may be made simpler because of it.

However, the differential equations that describe the dynamics of a system of particles are definitely nonlinear and so one must be somewhat more clever in applying the concept of perturbation theory. In this regard, celestial mechanics is a poor field in which to learn perturbation theory. One would be better served learning from a linear theory like quantum mechanics. Nevertheless, celestial mechanics is where we are, so we will make the best of it. Let us begin with a general statement of the approach for a conservative perturbing force.

9.1 The Basic Approach to the Perturbed Two Body Problem

The first step in any application of perturbation theory is to identify the space in which the perturbations are to be carried out and what variables are to be perturbed. At first glance, one could say that the ultimate result is to predict the position and velocity of one object with respect to another. Thus, one is tempted to look directly for perturbations to $\vec{r}$ as a function of time. However, the non-linearity of the equations of motion will make such an approach unworkable. Instead, let us make use of what we know about the solution to the two body problem.

For the two body problem we saw that the equations of motion have the form

$$\ddot{\vec{r}} + \nabla\Phi = 0 \quad, \tag{9.1.1}$$

where Φ is the potential of a point mass given by

$$\Phi = - GM/r \quad. \tag{9.1.2}$$

Let us assume that there is an additional source of a potential that can be represented by a scalar $-\Psi$ that introduces small forces acting on the object so that

$$|\nabla\Phi| \gg |\nabla\Psi| \quad. \tag{9.1.3}$$

We can then write the equations of motion as

$$\ddot{\vec{r}} + \nabla\Phi = \nabla\Psi(\vec{r},t) \quad. \tag{9.1.4}$$

Here Ψ is the negative of the perturbing potential by convention. If Ψ is a constant, then the solution to the equations of motion will be the solution to the two body problem. However, we already know that this will be a conic section which can be represented by six constants called the orbital elements. We also know that these six orbital elements can be divided into two triplets, the first of which deals with the size and shape of the orbit, and the second of which deals with the orientation of that orbit with respect to a specified coordinate system. A very reasonable question to ask is how does the presence of the perturbing potential affect the orbital elements. Clearly they will no longer be constants, but will vary in time. However, the knowledge of those constants as a function of time will allow us to predict the position and velocity of the object as well as its apparent location in the sky using the development in chapters 5, 6, and 7. This results from the fact that at any instant in time the object can be viewed as following an orbit that is a conic section. Only the characteristics of that conic section will be changing in time. Thus, the solution space appropriate for the perturbation analysis becomes the space defined by the six linearly independent orbital elements. That we can indeed do this results from the fact that the uniform motion of the center of mass provides the remaining six constants of integration even in a system of N bodies. Thus the determination of the temporal behavior of the orbital elements provides the remaining six pieces of linearly independent information required to uniquely determine the object's motion. The choice of the orbital elements as the set of parameters to perturb allows us to use all of the development of the two body problem to complete the solution.

Thus, let us define a vector $\vec{\xi}$ whose components are the instantaneous elements of the orbit so that we may regard the solution to the problem as given by

$$\vec{r} = \vec{r}(\vec{\xi}, t) \quad . \tag{9.1.5}$$

The problem has now been changed to finding how the orbital elements change in time due to the presence of the perturbing potential $-\Psi$. Explicitly we wish to recast the equations of motion as equations for $d\vec{\xi}/dt$. If we consider the case where the perturbing potential is zero, then $\vec{\xi}$ is constant so that we can write the unperturbed velocity as

$$\vec{v} = \frac{d\vec{r}}{dt} = \frac{\partial \vec{r}(\vec{\xi}, t)}{\partial t} \quad . \tag{9.1.6}$$

Now let us define a specific set of orbital elements $\vec{\xi}_0$ to be those that would determine the particle's motion if the perturbing potential suddenly became zero at some time t_0

$$\vec{\xi}_0 \equiv \vec{\xi}(t_0) \quad . \tag{9.1.7}$$

The orbital elements $\vec{\xi}_0$ represent an orbit that is tangent to the perturbed orbit at t_0 and is usually called the *osculating*

131

orbit. By the chain rule

$$\frac{d\vec{r}}{dt} = \frac{\partial\vec{r}}{\partial t} + \frac{\partial\vec{r}}{\partial\vec{\xi}}.\dot{\vec{\xi}} \qquad\qquad (9.1.8)$$

If we compare this with equation (9.1.6), it is clear that

$$\frac{\partial\vec{r}}{\partial\vec{\xi}}.\dot{\vec{\xi}} = 0 \qquad\qquad (9.1.9)$$

This is sometimes called the osculation condition. Applying this condition and differentiating equation (9.1.8) again with respect to time we get

$$\frac{d^2\vec{r}}{dt^2} = \frac{\partial^2\vec{r}}{\partial t^2} + \left(\frac{\partial}{\partial\vec{\xi}}\frac{\partial\vec{r}}{\partial t}\right).\dot{\vec{\xi}} \qquad\qquad (9.1.10)$$

If we replace (d^2r/dt^2) in equation (9.1.4) by equation (9.1.10), we get

$$\frac{\partial^2\vec{r}}{\partial t^2} + \nabla\Phi + \left(\frac{\partial}{\partial\vec{\xi}}\frac{\partial\vec{r}}{\partial t}\right).\dot{\vec{\xi}} = \nabla\Psi(\vec{r},t) \qquad\qquad (9.1.11)$$

However the explicit time dependence of $\vec{r}(\vec{\xi},t)$ is the same as $\vec{r}(\vec{\xi}_0,t)$ so that

$$\frac{\partial^2\vec{r}}{\partial t^2} + \nabla\Phi = 0 \qquad\qquad (9.1.12)$$

Therefore the first two terms of equation (9.1.11) sum to zero and, together with the osculating condition of equation (9.1.9), we have

$$\left. \begin{array}{l} \dot{\vec{\xi}}.\dfrac{\partial}{\partial\vec{\xi}}\left[\dfrac{\partial\vec{r}}{\partial t}\right] = \nabla\Psi \\[2em] \dot{\vec{\xi}}.\dfrac{\partial\vec{r}}{\partial\vec{\xi}} = 0 \end{array} \right\} \qquad\qquad (9.1.13)$$

as equations of condition for $\vec{\xi}$. Differentiation with respect to a vector simply means that the differentiation is carried out with respect to each component of the vector. Therefore

132

$[\partial \vec{r}/\partial \vec{\xi}]$ is a second rank tensor with components $[\partial r_i/\partial \xi_j]$. Thus each of the equations (9.1.13) are vector equations, so there is a separate scalar equation for each component of $\vec{r}$. Together they represent six nonlinear inhomogeneous partial differential equations for the six components of $\vec{\xi}$. The initial conditions for the solution are simply the values of $\vec{\xi}_0$ and its time derivatives at t_0. Appropriate mathematical rigor can be applied to find the conditions under which this system of equations will have a unique solution and this will happen as long as the Jacobian of $|\partial(\vec{r},\vec{v})/\partial \vec{\xi}| \neq 0$. Complete as these equations are, their form and application are something less than clear, so let us turn to a more specific application.

9.2 The Cartesian Formulation, Lagrangian Brackets, and Specific Formulae

Let us begin by writing equations (9.1.13) in component form. Assume that the cartesian components of $\vec{r}$ are x_i. Then equations (9.1.13) become

$$
\left.
\begin{aligned}
&\sum_{j=1}^{6} \frac{\partial \dot{x}_i}{\partial \xi_j} \frac{d\xi_i}{dt} = \frac{\partial \Psi}{\partial x_i} \qquad i = 1,2,3 \\[2em]
&\sum_{j=1}^{6} \frac{\partial x_i}{\partial \xi_j} \frac{d\xi_j}{dt} = 0 \qquad i = 1,2,3
\end{aligned}
\right\} \qquad (9.2.1)
$$

However, the dependence of ξ_j on time is buried in these equations and it would be useful to be able to write them so that $d\vec{\xi}/dt$ is explicitly displayed. To accomplish this multiply each of the first set of equations by $(\partial x_i/\partial \xi_k)$ and add the three component equations together. This yields

$$
\sum_{i=1}^{3} \frac{\partial x_i}{\partial \xi_k} \sum_{j=1}^{6} \frac{\partial \dot{x}_i}{\partial \xi_j} \frac{d\xi_j}{dt} = \sum_{i=1}^{3} \frac{\partial x_i}{\partial \xi_k} \frac{\partial \Psi}{\partial x_i} = \frac{\partial \Psi}{\partial \xi_k} \qquad k = 1,2,\cdots,6 \qquad (9.2.2)
$$

Multiply each of the second set of equations by $(-\partial \dot{x}_i/\partial \xi_k)$ and add them together to get

133

$$-\sum_{i=1}^{3} \frac{\partial \dot{x}_i}{\partial \xi_k} \sum_{j=1}^{6} \frac{\partial x_i}{\partial \xi_j} \frac{d\xi_j}{dt} = 0 \qquad k = 1,2,\cdots,6 \qquad (9.2.3)$$

Finally add equation (9.2.2) and equation (9.2.3) together, re-arrange the order of summation factoring out the desired quantity ($d\xi_j/dt$) to get

$$\sum_{j=1}^{6} \left(\frac{d\xi_j}{dt} \right) \sum_{i=1}^{3} \left[\frac{\partial x_i}{\partial \xi_k} \frac{\partial \dot{x}_i}{\partial \xi_j} - \frac{\partial x_i}{\partial \xi_j} \frac{\partial \dot{x}_i}{\partial \xi_k} \right] = \frac{\partial \Psi}{\partial \xi_k}, \quad k=1,2,\cdots,6 \quad (9.2.4)$$

The ugly looking term under the second summation sign is known as the *Lagrangian bracket* of ξ_k and ξ_j and, by convention, is written as

$$[\xi_k,\xi_j] \equiv \sum_{i=1}^{3} \left[\frac{\partial x_i}{\partial \xi_k} \frac{\partial \dot{x}_i}{\partial \xi_j} - \frac{\partial x_i}{\partial \xi_j} \frac{\partial \dot{x}_i}{\partial \xi_k} \right] \qquad (9.2.5)$$

The reason for pursuing this apparently complicating procedure is that the Lagrangian brackets have no explicit time dependence so that they represent a set of coefficients that simply multiply the time derivatives of ξ_j. This reduces the equations of motion to six first order linear differential equations which are

$$\sum_{j=1}^{6} [\xi_j,\xi_k] \frac{d\xi_j}{dt} = \frac{\partial \Psi}{\partial \xi_k} \qquad k = 1,2,\cdots,6 \qquad (9.2.6)$$

All we need to do is determine the Lagrangian brackets for an explicit set of orbital elements and since they are time independent, they may be evaluated at any convenient time such as t_0.

   If we require that the scalar (dot) product be taken over coordinate ($\vec{r}$) space rather than orbital element ($\vec{\xi}$) space, we can write the Lagrangian bracket as

$$[\xi_k,\xi_j] \equiv [\vec{\xi},\vec{\xi}] = \left[\frac{\partial \vec{r}}{\partial \vec{\xi}} \cdot \frac{\partial \dot{\vec{r}}}{\partial \vec{\xi}} - \frac{\partial \dot{\vec{r}}}{\partial \vec{\xi}} \cdot \frac{\partial \vec{r}}{\partial \vec{\xi}} \right] \qquad (9.2.7)$$

134

Since the partial derivatives are tensors, the scalar product in coordinate space does not commute. However, we may show the lack of explicit time dependence of the Lagrange bracket by direct partial differentiation with respect to time so that

$$\frac{\partial}{\partial t}[\vec{\xi},\vec{\xi}] = \left[\frac{\partial^2\vec{r}}{\partial t\partial\vec{\xi}}\cdot\frac{\partial\dot{\vec{r}}}{\partial\vec{\xi}} + \frac{\partial\vec{r}}{\partial\vec{\xi}}\cdot\frac{\partial^2\dot{\vec{r}}}{\partial t\partial\vec{\xi}}\right] - \left[\frac{\partial^2\dot{\vec{r}}}{\partial t\partial\vec{\xi}}\cdot\frac{\partial\vec{r}}{\partial\vec{\xi}} + \frac{\partial\dot{\vec{r}}}{\partial\vec{\xi}}\cdot\frac{\partial^2\vec{r}}{\partial t\partial\vec{\xi}}\right], \quad (9.2.8)$$

or re-arranging the order of differentiation we get

$$\frac{\partial}{\partial t}[\vec{\xi},\vec{\xi}] = \frac{\partial}{\partial\vec{\xi}}\left[\frac{\partial\vec{r}}{\partial t}\cdot\frac{\partial\dot{\vec{r}}}{\partial\vec{\xi}} - \frac{\partial\vec{r}}{\partial\vec{\xi}}\cdot\frac{\partial\dot{\vec{r}}}{\partial t}\right] - \frac{\partial}{\partial\vec{\xi}}\left[\frac{\partial\dot{\vec{r}}}{\partial\vec{\xi}}\cdot\frac{\partial\vec{r}}{\partial t} - \frac{\partial\dot{\vec{r}}}{\partial\vec{\xi}}\cdot\frac{\partial\vec{r}}{\partial t}\right]. \quad (9.2.9)$$

Using equation (9.1.6) and Newton's laws we can write this as

$$\frac{\partial}{\partial t}[\vec{\xi},\vec{\xi}] = \frac{\partial}{\partial\vec{\xi}}\left[\frac{1}{2}\frac{\partial v^2}{\partial\vec{\xi}} - \frac{\partial\Phi}{\partial\vec{\xi}}\right] - \frac{\partial}{\partial\vec{\xi}}\left[\frac{1}{2}\frac{\partial v^2}{\partial\vec{\xi}} - \frac{\partial\Phi}{\partial\vec{r}}\cdot\frac{\partial\vec{r}}{\partial\vec{\xi}}\right] = 0 \quad (9.2.10)$$

Remember that we wrote $\vec{r}(\vec{\xi},t)$ so that the coordinates x_i and their time derivatives $\dot{x}_i$ depend only on the set of orbital elements ξ_j and time. Thus the Lagrangian brackets depend only on the particular set of orbital elements and may be computed once and for all. There are various procedures for doing this, some of which are tedious and some of which are clever, but all of which are relatively long. For example, one can calculate them for $t = T_0$ so that $M = E = \nu = 0$. In addition, while one can formulate 36 values of $[\xi_k,\xi_j]$ it is clear from equation (9.2.5) that

$$\left.\begin{array}{l} [\xi_k,\xi_j] = -[\xi_j,\xi_k] \\[6pt] [\xi_k,\xi_k] = 0 \end{array}\right\} \quad (9.2.11)$$

This reduces the number of linearly independent values of $[\xi_k,\xi_j]$ to 15. However, of these 15 Lagrange brackets, only 6 are nonzero [see Taff[11] p.306,307] and are given below.

$$\left.\begin{array}{l} [i,\Omega] = na^2(1-e^2)^{\frac{1}{2}}\sin(i) \\[6pt] [a,\Omega] = -\frac{1}{2}na(1-e^2)^{\frac{1}{2}}\cos(i) \\[6pt] [e,\Omega] = na^2e(1-e^2)^{-\frac{1}{2}}\cos(i) \\[6pt] [a,\omega] = -\frac{1}{2}na(1-e^2)^{\frac{1}{2}} \\[6pt] [e,\omega] = na^2e(1-e)^{-\frac{1}{2}} \end{array}\right\} \quad (9.2.12)$$

135

$$[a,T] = \tfrac{1}{2}a \qquad \Bigg\} \quad ,$$

where n is just the mean angular motion given in terms of the mean anomaly M by

$$M = n(t-T) \quad . \tag{9.2.13}$$

Thus the coefficients of the time derivatives of ξ_j are explicitly determined in terms of the orbital elements of the osculating orbit $\vec{\xi}_0$.

To complete the solution, we must deal with the right hand side of equation (9.2.6). Unfortunately, the partial derivatives of the perturbing potential are likely to involve the orbital elements in a complicated fashion. However, we must say something about the perturbing potential or the problem cannot be solved. Therefore, let us assume that the behavior of the potential is understood in a cylindrical coordinate frame with radial, azimuthal, and vertical coordinates $(r, \vartheta,$ and $z)$ respectively. We will then assume that the cylindrical components of the perturbing force are known and given by

$$\begin{aligned}
\mathfrak{R} &\equiv \frac{\partial \Psi}{\partial r} \\[2mm]
\mathfrak{I} &\equiv \frac{1}{r}\frac{\partial \Psi}{\partial \vartheta} \\[2mm]
\mathfrak{C} &\equiv \frac{\partial \Psi}{\partial z}
\end{aligned} \Bigg\} \tag{9.2.14}$$

Then from the chain rule

$$\frac{\partial \Psi}{\partial \xi_i} = \frac{\partial \Psi}{\partial r}\frac{\partial r}{\partial \xi_i} + \frac{\partial \Psi}{\partial \vartheta}\frac{\partial \vartheta}{\partial \xi_i} + \frac{\partial \Psi}{\partial z}\frac{\partial z}{\partial \xi_i} \quad . \tag{9.2.15}$$

The partial derivatives of the cylindrical coordinates with respect to the orbital elements may be calculated directly and the equations for the time derivatives of the orbital elements [equations (9.2.6)] solved explicitly. The algebra is long and tedious but relatively straight forward and one gets

$$\frac{da}{dt} = \frac{2}{n}(1-e^2)^{\frac{1}{2}}\{\Re e\ \sin\nu + \Im a(1-e^2)/r\}$$

$$\frac{de}{dt} = \frac{1}{na}(1-e^2)^{\frac{1}{2}}\{\Re\ \sin\nu + \Im[\cos(E) + \cos\nu]\}$$

$$\frac{di}{dt} = [na^2(1-e^2)^{\frac{1}{2}}]^{-1}\mathfrak{C}r\ \cos(\nu+\omega)$$

$$\frac{d\Omega}{dt} = [na^2(1-e^2)^{\frac{1}{2}}\sin(i)]^{-1}\mathfrak{C}r\ \sin(\nu+\omega)$$

$$\frac{d\omega}{dt} = \frac{(1-e^2)^{\frac{1}{2}}}{nae}\left[\Im\{\sin\nu[r+a(1-e^2)]/a(1-e^2)\} - \Re\cos\nu\right]$$

$$+ \frac{d\Omega}{dt}\left[2(1-e^2)^{\frac{1}{2}}\sin^2(i/2) - 1\right]$$

$$\frac{dT}{dt} = \frac{1}{n}\left\{\frac{d\Omega}{dt}\left[1+2(1-e^2)^{\frac{1}{2}}\sin^2(i/2)\right]\right.$$

$$- \left[\frac{d\Omega}{dt} + \frac{d\omega}{dt}\right]e^2[1+(1-e^2)^{\frac{1}{2}}]^{-1}$$

$$\left. - [2r\Re/na^2]\right\}$$

(9.2.16)

An alternative set of perturbation equations attributed to Gauss and given by Taff[11] (p.314) is

$$\frac{da}{dt} = \frac{2e\ \sin\nu}{n(1-e^2)^{\frac{1}{2}}}\Re + \frac{2a(1-e^2)^{\frac{1}{2}}}{nr}\Im$$

$$\frac{de}{dt} = \frac{(1-e^2)^{\frac{1}{2}}\sin\nu}{na}\Re + \frac{(1-e^2)^{\frac{1}{2}}}{na^2e}\left[\frac{a^2(1-e^2)-r^2}{r^2}\right]\Im$$

(9.2.17)

137

$$\frac{d\omega}{dt} = -\frac{(1-e^2)^{\frac{1}{2}}\cos\nu}{nae}\mathfrak{R} + \frac{(1-e^2)^{\frac{1}{2}}\sin\nu}{nae}\left[\frac{a(1-e^2)+r}{a(1-e^2)}\right]\mathfrak{S}$$

$$-\frac{r\,\sin(\Omega+\omega)\cot(i)}{L}\mathfrak{C}$$

$$\frac{di}{dt} = \frac{r\,\cos(\Omega+\omega)}{L}\mathfrak{C}$$

$$\frac{d\Omega}{dt} = \frac{r\,\csc(i)\,\sin(\Omega+\omega)}{L}\mathfrak{C}$$

$$\frac{dT}{dt} = +\frac{1}{n^2a}\left[\frac{2r}{a} - \frac{(1-e^2)\cos\nu}{e}\right]\mathfrak{R} + \frac{(1-e^2)\sin\nu}{na^2e}\left[\frac{a(1-e^2)+r}{a(1-e^2)}\right]\mathfrak{S},$$

where

$$L = 2\pi a^2/P \qquad . \tag{9.2.18}$$

These relatively complicated forms for the solution show the degree of complexity introduced by the nonlinearity of the equations of motion. However, they are sufficient to demonstrate that the problem does indeed have a solution. Given the perturbing potential and an approximate two body solution at some epoch t_0, one can use all of the two body mechanics developed in previous chapters to calculate the quantities on the right hand side of equations (9.2.16). This allows for a new set of orbital elements to be calculated and the motion of the objects followed in time. The process may be repeated allowing for the cumulative effects of the perturbation to be included.

However, one usually relies on the original assumption that the perturbing forces are small compared to those that produce the two body motion [equation (9.1.3)]. Then all the terms on the left hand side of equation (9.2.16) will be small and the motion can be followed for many orbits before it is necessary to change the orbital elements. That is the major thrust of perturbation theory. It tells you how things ought to change in response to known forces. Thus, if the source of the perturbation lies in the plane defining the cylindrical coordinate system (and the plane defining the orbital elements) $\mathfrak{C}$ will be zero and the orbital inclination i and the longitude of the ascending node will not change in time. Similarly if the source lies along the z-axis of the system, the semi-major axis

138

(a) and eccentricity (e) will be time independent. If the changes in the orbital elements are sufficiently small so that one may average over an orbit without any significant change, then many of the perturbations vanish. In any event, such an averaging procedure may be used to determine equations for the slow change of the orbital elements.

Utility of the development of these perturbation equations relies on the approximation made in equation (9.1.3). That is, the equations are essentially first order in the perturbing potential. Attempts to include higher order terms have generally led to disaster. The problem is basically that the equations of classical mechanics are nonlinear and that the object of interest is $\vec{r}(t)$. Many small errors can propagate through the procedures for finding the orbital elements and then to the position vector itself. Since the equations are nonlinear, the propagation is nonlinear. In general, perturbation theory has not been terribly successful in solving problems of celestial mechanics. So the current approach is generally to solve the Newtonian equations of motion directly using numerical techniques. Awkward as this approach is, it has had great success in solving specific problems as is evidenced by the space program. The ability to send a rocket on a complicated trajectory through the satellite system of Jupiter is ample proof of that. However, one gains little general insight into the effects of perturbing potentials from single numerical solutions.

Problems such as the Kirkwood Gaps and the structure of the Saturnian ring system offer ample evidence of problems that remain unsolved by classical celestial mechanics. However, in the case of the former, much light has been shed through the dynamics of Chaos (see Wisdom[10]). There remains much to be solved in celestial mechanics and the basic nonlinearity of the equations of motion will guarantee that the solutions will not come easily.

Formal perturbation theory provides a nice adjunct to the formal theory of celestial mechanics as it shows the potential power of various techniques of classical mechanics in dealing with problems of orbital motion. Because of the nonlinearity of the Newtonian equations of motion, the solution to even the simplest problem can become very involved. Nevertheless, the majority of dynamics problems involving a few objects can be solved one way or another. Perhaps it is because of this non-linearity that so many different areas of mathematics and physics must be brought together in order to solve these problems. At any rate celestial mechanics provides a challenging training field for students of mathematical physics to apply what they know.

Chapter 9: Exercises

1. If the semi-major axis of a planets orbit is changed by Δa, how does the period change? How does a change in the orbital eccentricity Δe affect the period?

2. If v_1 and v_2 are the velocities of a planet at perihelion and aphelion respectively, show that

$$(1-e)v_1 = (1+e)v_2 \quad .$$

3. Find the Lagrangian bracket for $[e, \Omega]$.

4. Using the Lagrangian and Gaussian perturbation equations, find the behavior of the orbital elements for a perturbative potential that has a pure r-dependence and is located at the origin of the coordinate system.

References and Supplemental Reading

1. Morse,P.M., and Feshbach, H., "Methods of Theoretical Physics" (1953) McGraw-Hill Book Company, Inc., New York, Toronto, London pp655-666.

2. Goldstein, H., "Classical Mechanics" (1959) Addison-Wesley Pub. Co. Inc. Reading, London

3. "Astronomical Almanac for the Year 1988" (1987) U.S. Government Printing Office, Washington

4. Jackson, J.D., "Classical Electrodynamics" (1962) John Wiley & Sons, New York, London, Sydney, pp.98-131.

5. Brouwer, D., and Clemence, G.M., "Methods of Celestial Mechanics" (1961) Academic Press, New York and London

6. Green, R.M., "Spherical Astronomy" (1985) Cambridge University Press, Cambridge pp144-147

7. Danby, J.M.A., "Fundamentals of Celestial Mechanics" (1962) The Macmillan Company, New York

8. Moulton, F.R., "An Introduction to Celestial Mechanics" 2nd Rev Ed. (1914) The Macmillan Company, New York

9. Collins, G.W.,II, "The Virial Theorem in Stellar Astrophysics" (1978) Pachart Publishing House, Tucson

10. Wisdom, J. "Urey Prize Lecture: Chaotic Dynamics in the Solar System" (1987) Icarus 72 pp241-275.

11. Taff,L.G., "Celestial Mechanics: A computational guide for the practioner" (1985) John Wiley & Sons

The references above constitute required reading for any who would become a practioner of celestial mechanics. Certainly Morse and Feshbach is one of the most venerable texts on theoretical physics and contains more information than most theoreticians would use in a lifetime. However, the book should be in the arsenal that any theoretician brings to the problems of analysis in physics. I still feel that Goldstein's text on classical mechanics is the best and most complete of the current era. However, some may find the text by Symon somewhat less condensed. The text by Brouwer and Clemence is the most advanced of the current texts in the field of celestial mechanics and is liable to remain so for some time to come. It is rather formidable, but contains information on such a wide range of problems and techniques that it should be at least perused by any student of the field. The text by Danby was the logical successor to the time honored work of Moulton. Danby introduced vector notation to the subject and made the reading much simpler. A.E.Roy expanded on this approach and covered a much wider range of topics. The celestial mechanics text by Fitzgerald listed below provides a development more common to modern day celestial mechanics and contains an emphasis on the orbital mechanics of satellites. This point of view is also used by Escobal where the first book on the "Methods of Orbit Determination" lays the groundwork for a contemporary discussion of 'rocket navigation' in the second book on "Astrodynamics". A much broader view of the term astrodynamics is taken by Herrick in his two volume treatise on the subject. The five volume 'epic' by Hagihara tries to summarize all that has happened in celestial mechanics in the last century and comes close to doing so. The text by Taff is one of the most recent of the celestial mechanics texts mentioned here, but still largely follows the traditional development started by Moulton. The exception is his discussion of perturbation theory which I found philosophically satisfying. The Urey Prize lecture by Wisdom should be read in its entirety by anyone who is interested in the application of the mathematics of Chaos to objects in the solar system.

Below I have given some additional references as 'supplemental reading' which I have found helpful from time to time in dealing with the material covered in this book. Most any book on modern algebra will contain definitions of what constitutes a set or group, but I found Andree very clear and concise. One of the best all round books on mathematical analysis with a view to numerical applications is that by Arfken. It is remarkably complete and wide ranging. The two articles from Chaotic Phenomena in Astrophysics show some further application of the subjects discussed by Wisdom. However, the entire book is interesting as it demonstrates how this developing field of mathematics has found applications in a number of areas of astrophysics.

Sokolnikoff and Redheffer is just one of those omnibus references that provides a myriad of definitions and development for mathematical analysis necessary for any student of the physical sciences. On the other hand, the lectures by Ogorodnikov provide one of the most lucid accounts of Liouville's Theorem and the implications for a dynamical system in phase space. The text on Gravitation by Misner, Thorne, and Wheeler has probably the most contemporary and complete treatment of tensors as they apply to the physical world. Although the main subject is somewhat tangent to celestial mechanics, it is a book that every educated physicist or astrophysicist must read. Since it is rather long, one should begin early. One should not leave the references of celestial mechanics without a mention of the rare monograph by Paul Herget. While the presentation of the material is somewhat encumbered by numerical calculations for which Paul Herget was justly renowned, the clarity of his understanding of the problems of classical orbit calculation makes reading this work most worthwhile.

1. Andree, R.V. "Selections from Modern Abstract Algebra" (1958) Henry Holt and Co.,
                                                                        New York

2. Arfken, G. "Mathematical Methods for Physicists" 2nd ed. (1970) Academic Press,
                                                    New York, San Francisco, London

3. Bensimon, D. and Kadanoff, L.P., "The Breakdown of KAM Trajectories" in "Chaotic Phenomena in Astrophysics" (1987) Ed H. Eichhorn and J.R. Buchler
                                        Ann. New York Acad. Sci. 497, pp110-117.

4. Escobal, P.R., "Methods of Orbit Determination" (1965) John Wiley and Sons, Inc., New
                                                        York, London, Sydney

5. _____, "Methods of Astrodynamics" (1968) John Wiley and Sons, Inc., New York,
                                                            London, Sydney

6. Fitzpatrick, P.M., "Principles of Celestial Mechanics" (1970)
                                            Academic Press Inc, New York, London

7. Hagihara, Y. "Celestial Mechanics" Vol. 1-5 (1970-1972)   MIT Press, Cambridge Mass.

8. Herget, P., "The Computation of Orbits" (1948)   Privately published by the author.

9. Herrick, S., "Astrodynamics" Vol. 1. (1971)   Van Nostrand Reinhold Company, London

10. _____, "Astrodynamics" Vol. 2. (1972)   Van Nostrand Reinhold Company, London

11. Meiss, J.D., "Resonances Fill Stochastic Phase Space" in "Chaotic Phenomena in Astrophysics" (1987) Ed. H. Eichhorn and J.R. Buchler
                                        Ann. New York Acad. Sci. 497, pp83-96.

12. Misner, C.W., Thorne, K.S., and Wheeler, J.A., "Gravitation" (1973)
                                            W.H.Freeman and Co. San Francisco

13. Ogorodnikov, K.F. "Dynamics of Stellar Systems" (1965) Trans. J.B.Sykes
                                    ed. A. Beer The Macmillian Company, New York

14. Roy, A.E., "Orbital Motion" (1982)   Adam Hilger Ltd., Bristol

15. Sokolnikoff, I.S., and Redheffer, R.M. "Mathematics of Physics and Modern Engineering" (1958)   McGraw Hill Book Co. Inc., New York, Toronto, London

16. Symon, K.R. "Mechanics" (1953)   Addison-Wesley Pub. Co. Inc., Reading

Index

A